Building the U.S. Battery Industry for Electric Drive Vehicles

PROGRESS, CHALLENGES, AND OPPORTUNITIES

Summary of a Symposium

Charles W. Wessner, Rappor

Subcommittee on Electric Drive Battery Research and Development Activities

Committee on Competing in the 21st Century:
Best Practice in State and Regional Innovation Initiatives

Board on Science, Technology, and Economic Policy

Policy and Global Affairs

NATIONAL RESEARCH COUNCIL
OF THE NATIONAL ACADEMIES

THE NATIONAL ACADEMIES PRESS
Washington, D.C.
www.nap.edu

THE NATIONAL ACADEMIES PRESS 500 Fifth Street NW Washington DC 20001

NOTICE: The project that is the subject of this report was approved by the Governing Board the National Research Council, whose members are drawn from the councils of the National Academy of Sciences, the National Academy of Engineering, and the Institute of Medicine. T members of the committee responsible for the report were chosen for their special competenc and with regard for appropriate balance.

This study was supported by: Contract/Grant No. DE-DT0000584, TO# 29, between the National Academy of Sciences and the Department of Energy. This report was prepared by the National Academy of Sciences under award number SB134106Z0011, TO# 4 (68059) from the U.S. Department of Commerce, National Institute of Standards and Technology (NIST). This report was prepared by the National Academy of Sciences under award number 99-06-07543-02 from the Economic Development Administration, U.S. Department of Commerce. The statements, findings, conclusions, and recommendations are those of the author(s) and do not necessarily reflect the views of the National Institute of Standards and Technology, the Economic Development Administration, or the U.S. Department of Commerce. Additional support was provided by the Michigan Economic Development Corporation and Michigan's University Research Corridor. Any opinions, findings, conclusions, or recommendations expressed in this publication are those of the author(s) and do not necessarily reflect the views of the organizations or agencies that provided support for the project.

International Standard Book Number 13: 978-0-309-25452-6 (Book)
International Standard Book Number 10: 0-309-25452-3 (Book)

Additional copies of this report are available for sale from the National Academies Press, 500 Fifth Street, N.W., Keck 360, Washington, DC 20001; (800) 624-6242 or (202) 334-3313; http://www.nap.edu/.

Printed in the United States of America

THE NATIONAL ACADEMIES

Advisers to the Nation on Science, Engineering, and Medicine

The **National Academy of Sciences** is a private, nonprofit, self-perpetuating society of distinguished scholars engaged in scientific and engineering research, dedicated to the furtherance of science and technology and to their use for the general welfare. Upon the authority of the charter granted to it by the Congress in 1863, the Academy has a mandate that requires it to advise the federal government on scientific and technical matters. Dr. Ralph J. Cicerone is president of the National Academy of Sciences.

The **National Academy of Engineering** was established in 1964, under the charter of the National Academy of Sciences, as a parallel organization of outstanding engineers. It is autonomous in its administration and in the selection of its members, sharing with the National Academy of Sciences the responsibility for advising the federal government. The National Academy of Engineering also sponsors engineering programs aimed at meeting national needs, encourages education and research, and recognizes the superior achievements of engineers. Dr. Charles M. Vest is president of the National Academy of Engineering.

The **Institute of Medicine** was established in 1970 by the National Academy of Sciences to secure the services of eminent members of appropriate professions in the examination of policy matters pertaining to the health of the public. The Institute acts under the responsibility given to the National Academy of Sciences by its congressional charter to be an adviser to the federal government and, upon its own initiative, to identify issues of medical care, research, and education. Dr. Harvey V. Fineberg is president of the Institute of Medicine.

The **National Research Council** was organized by the National Academy of Sciences in 1916 to associate the broad community of science and technology with the Academy's purposes of furthering knowledge and advising the federal government. Functioning in accordance with general policies determined by the Academy, the Council has become the principal operating agency of both the National Academy of Sciences and the National Academy of Engineering in providing services to the government, the public, and the scientific and engineering communities. The Council is administered jointly by both Academies and the Institute of Medicine. Dr. Ralph J. Cicerone and Dr. Charles M. Vest are chair and vice chair, respectively, of the National Research Council.

www.national-academies.org

Subcommittee on
Electric Drive Battery Research and Development Activities

Mary L. Good (NAE), *Chair*
Dean Emeritus, Donaghey College of Engineering and Information Technology
Special Advisor to the Chancellor for Economic Development
University of Arkansas at Little Rock

Raymond G. Boeman
Director, Energy Partnerships
Oak Ridge National Laboratory

Michael G. Borrus
Founding General Partner
X/Seed Capital Management

Ralph Brodd
President
Broddarp of Nevada

Robert Kruse
Principal
EV Consulting

W. Clark McFadden II
Senior Counsel
Orrick, Herrington & Sutcliffe LLP

Daniel Sperling
Director, Institute of Transportation Studies
University of California, Davis

Committee on Competing in the 21st Century:
Best Practice in State and Regional Innovation Initiatives

Mary L. Good (NAE), ***Chair***
Dean Emeritus, Donaghey College of Engineering and Information Technology
Special Advisor to the Chancellor for Economic Development
University of Arkansas at Little Rock

Michael G. Borrus
Founding General Partner
X/Seed Capital Management

William C. Harris
President and CEO
Science Foundation Arizona

W. Clark McFadden II
Senior Counsel
Orrick, Herrington & Sutcliffe LLP

David T. Morgenthaler
Founding Partner
Morgenthaler Ventures

Edward E. Penhoet (IOM)
Director
Alta Partners

Tyrone C. Taylor
President
Capitol Advisors on Technology, LLC

PROJECT STAFF

Charles W. Wessner
Study Director

McAlister T. Clabaugh
Program Officer

David S. Dawson
Senior Program Assistant

Sujai J. Shivakumar
Senior Program Officer

David E. Dierksheide
Program Officer

Peter Engardio
Consultant

For the National Research Council (NRC), this project was overseen by the Board on Science, Technology and Economic Policy (STEP), a standing board of the NRC established by the National Academies of Sciences and Engineering and the Institute of Medicine in 1991. The mandate of the Board on Science, Technology, and Economic Policy is to advise federal, state, and local governments and inform the public about economic and related public policies to promote the creation, diffusion, and application of new scientific and technical knowledge to enhance the productivity and competitiveness of the U.S. economy and foster economic prosperity for all Americans. The STEP Board and its committees marshal research and the expertise of scholars, industrial managers, investors, and former public officials in a wide range of policy areas that affect the speed and direction of scientific and technological change and their contributions to the growth of the U.S. and global economies. Results are communicated through reports, conferences, workshops, briefings, and electronic media subject to the procedures of the National Academies to ensure their authoritativeness, independence, and objectivity. The members of the STEP Board* and the NRC staff are listed below:

*As of September 2012.

continued

STEP Staff

Contents

Preface

Responding to the challenges of fostering regional growth and employment in an increasingly competitive global economy, many U.S. states and regions have developed programs to attract and grow high-technology companies, develop the talent and resources necessary to create innovation clusters, and sustain manufacturing and high value employment. These state and regionally based initiatives have a broad range of goals and increasingly include significant resources that often focus on driving innovation and often in partnership with foundations and universities. These are being joined by recent initiatives to coordinate and concentrate investments from a variety of federal agencies that provide significant resources to develop regional centers of innovation, business incubators, and other strategies to encourage entrepreneurship and high-tech development.

In this regard, the state of Michigan is making significant investments to develop an electrified-vehicle industrial cluster. The state offered more than $1 billion in grants and tax credits to manufacturers of lithium-ion battery cells, packs, and components. Michigan has also invested in research centers and skilled-worker training programs for electrified vehicles.

Efforts by the federal government to ensure that the U.S. has a domestic manufacturing base for advanced batteries are complementing Michigan's initiatives. The federal government in 2009 awarded $2.4 billion in grants under the American Recovery and Reinvestment Act to manufacturers of lithium-ion cells, battery packs, and materials.[1] A host of other financial incentives have also been introduced to help companies commercialize new vehicle technologies, build production lines, build supply chains, and encourage consumers to buy electric-gas hybrid cars.

To review the developments, as well as the needs and challenges, of the U.S. electric drive battery industry in Michigan, the National Academies Board on Science, Technology, and Economic Policy (STEP), in cooperation with the Michigan Economic Development Corporation and the Department of Energy, convened a symposium, on *Building the U.S. Battery Industry for Electric Drive Vehicles: Progress, Challenges, and Opportunities.*

The symposium, held on 26-27 July 2010 in Livonia, Michigan, and this report of that symposium, address the first of two key elements of the Statement of Task (described below) of a committee of the National Research Council.

---

[1] The American Recovery and Reinvestment Act of 2009 (P. L. 115-5) is a $787 billion economic stimulus packaged signed by President Barack Obama on Feb. 17, 2009. See Department of Energy, "The Recovery Act: Transforming America's Transportation Sector—Batteries and Electric Vehicles," July 14, 2010 (http://www.whitehouse.gov/files/documents/Battery-and-Electric-Vehicle-Report-FINAL.pdf)

STATEMENT OF TASK

The Overall Project

An ad hoc subcommittee will plan and conduct two public symposia to review and analyze the potential contributions of public-private partnerships and identify other relevant issues for the Department of Energy, Office of Vehicle Technologies, Energy Storage Team's activities in the energy storage research and development area. The symposia will also identify lessons from these and other domestic and international experiences to help inform DoE as to whether its activities are complete and appropriately focused. Additional topics that emerge in the course of the planning may also be addressed. The two symposia will gather representatives from leading battery manufacturers, automotive firms, university researchers, academic and industry analysts, congressional staff, and federal agency representatives. An individually-authored summary of each symposium will be issued.

This Report

The symposium that is the subject of this report was held in Michigan in order to provide direct access to the policymakers and industrial participants drawn from the concentration of battery manufacturers and automotive firms in the region. The symposium reviewed the current state, needs, and challenges of the U.S. advanced battery manufacturing industry; challenges and opportunities in battery R&D, commercialization, and deployment; collaborations between the automotive industry and battery industry; workforce issues, and supply chain development. It also focused on the impact of DoE's investments and the role of state and federal programs in support of this growing industry. This task of this report is to summarize the presentations and discussions that took place at this symposium. Needless to say, the battery industry has evolved very substantially since the conference was held, and indeed some of the caveats raised by the speakers with regard to overall demand for batteries and the prospects of multiple producers now seem prescient. At the same time, it is important to understand that it is unrealistic to expect that all recipients of local, state, or federal support in a complex and rapidly evolving industry will necessarily succeed. A number of the firms discussed here have been absorbed by competitors, others have gone out of business, and others continue to progress.[2]

[2] The Overview chapter of this report takes note of these recent developments.

THE CONTEXT OF THIS REPORT

Since 1991, the National Research Council, under the auspices of the Board on Science, Technology, and Economic Policy, has undertaken a program of activities to improve policymakers' understandings of the interconnections of science, technology, and economic policy and their importance for the American economy and its international competitive position. The Board's activities have corresponded with increased policy recognition of the importance of knowledge and technology to economic growth.

One important element of STEP's analysis concerns the growth and impact of foreign technology programs.[3] U.S. competitors have launched substantial programs to support new technologies, small firm development, and consortia among large and small firms to strengthen national and regional positions in strategic sectors. Some governments overseas have chosen to provide public support to innovation to overcome the market imperfections apparent in their national innovation systems.[4] They believe that the rising costs and risks associated with new potentially high-payoff technologies, and the growing global dispersal of technical expertise, underscore the need for national R&D programs to support new and existing high-technology firms within their borders.

Similarly, many state and local governments and regional entities in the United States are undertaking a variety of initiatives to enhance local economic development and employment through investment programs designed to attract knowledge-based industries and grow innovation clusters.[5] These state and regional programs and associated policy measures are of great interest for their potential contributions to growth and U.S. competitiveness and for the "best practice" lessons they offer for other state and regional programs.

STEP's project on State and Regional Innovation Initiatives is intended to generate a better understanding of the challenges associated with the transition of research into products, the practices associated with successful state and regional programs, and their interaction with federal programs and private initiatives. The study seeks to achieve this goal through a series of complementary assessments of state, regional, and federal initiatives; analyses of specific industries and technologies from the perspective of crafting supportive public policy at all three levels; and outreach to multiple

---

[3] National Research Council, *Innovation Policies for the 21st Century*, C. Wessner, ed., Washington DC: National Academies Press, 2007.

[4] For example, a number of countries are investing significant funds in the development of research parks. For a review of selected national efforts, see National Research Council, *Understanding Research, Science and Technology Parks: Global Best Practices*, C. Wessner, ed., Washington, DC: National Academies Press, 2009.

[5] For a scoreboard of state efforts, see Robert Atkinson and Scott Andes, *The 2010 State New Economy Index: Benchmarking Economic Transformation in the States,* Kauffman Foundation and ITIF, November 2010.

stakeholders. The overall goal is to improve the operation of state and regional programs and, collectively, enhance their impact.

WORKSHOP SUMMARY

This report captures the presentations and discussions of the STEP symposium on *Building the U.S. Battery Industry for Electric Drive Vehicles: Progress, Challenges, and Opportunities*. It includes an introduction highlighting key issues raised at the meeting and summary of the meeting's presentations. This workshop summary has been prepared by the workshop rapporteur as a factual summary of what occurred at the workshop. The planning committee's role was limited to planning and convening the workshop. The statements made are those of the rapporteur or individual workshop participants and do not necessarily represent the views of all workshop participants, the planning committee, or the National Academies.

ACKNOWLEDGMENTS

On behalf of the National Academies, we express our appreciation for the insights, expertise, and perspectives provided by the many well-informed contributors to this meeting. We would also like to extend special recognition to Gary Krause from the Michigan Economic Development Corporation and his colleagues, along with McAlister Clabaugh and David Dawson of the STEP staff, for their commitment and excellent organization of the event. We are also indebted to Peter Engardio, formerly of Businessweek and now with the Boston Consulting Group, for his preparation of the introduction and summary of the meeting. We also wish to thank Dr. Sujai Shivakumar and David Dawson of the STEP staff for their tireless efforts to prepare the report for publication among many other competing priorities.

NATIONAL RESEARCH COUNCIL REVIEW

This report has been reviewed in draft form by individuals chosen for their diverse perspectives and technical expertise, in accordance with procedures approved by the National Academies' Report Review Committee. The purpose of this independent review is to provide candid and critical comments that will assist the institution in making its published report as sound as possible and to ensure that the report meets institutional standards for quality and objectivity. The review comments and draft manuscript remain confidential to protect the integrity of the process.

We wish to thank the following individuals for their review of this report: Robert Bachrach, Energy & Environmental Solutions; Robert Boege, ASTRA; Martin Dober, Michigan Economic Development Corporation; and Paul DeCotis, Long Island Power Authority. Although the reviewers listed above have provided many constructive comments and suggestions, they were not asked to endorse the content of the report, nor did they see the final draft before its release. Responsibility for the final content of this report rests entirely with the STEP Board and the institution.

Charles W. Wessner Mary L. Good

I

OVERVIEW

Overview

Only a few years ago, the United States faced the prospect of entering the age of electrified transportation without a significant domestic advanced battery manufacturing industry. Virtually all lithium-ion battery cells, widely expected to be a core technology for electric cars and trucks of the future, were made in Asia. Even though there were many promising U.S. start-ups with innovative lithium-ion battery technology for cars, few could raise funds to build factories in America.

To address this gap and to ensure that the U.S. would have a domestic manufacturing base for advanced batteries, the federal government awarded $2.4 billion in grants in 2009 under the American Recovery and Reinvestment Act to manufacturers of lithium-ion cells, battery packs, and materials.[1] A host of other financial incentives were also introduced to help companies commercialize new vehicle technologies, build production lines, and encourage consumers to buy hybrid cars. These grants complemented the $25 billion in debt capital made available by the federal government to encourage automakers produce more energy-efficient cars under the Advanced Technology Vehicles Manufacturing (ATVM) Loan Program.[2]

The state of Michigan has also made significant investments to develop an electrified-vehicle industrial cluster. The state offered more than $1 billion in grants and tax credits to manufacturers of lithium-ion battery cells, packs, and components. Michigan also invested in research centers and skilled-worker training programs for electrified vehicles.

Based on these federal and state initiatives, some 16 battery-related factories were being built in Michigan as of mid-2010. These investments were projected to create 62,000 jobs in five years.[3] However, while Michigan and other states are now building substantial assembly capacity for advanced batteries, the nascent U.S. advanced battery industry remains in a "most critical state of development," as A123 Systems executive James M. Forcier has observed.[4] The core issue is whether there be enough demand for hybrid and electric vehicles to sustain the industry.[5] Another pressing question is whether

---

[1] The American Recovery and Reinvestment Act of 2009 (P. L. 115-5) is a $787 billion economic stimulus packaged signed by President Barack Obama on Feb. 17, 2009. See Department of Energy, "The Recovery Act: Transforming America's Transportation Sector—Batteries and Electric Vehicles," July 14, 2010 (http://www.whitehouse.gov/files/documents/Battery-and-Electric-Vehicle-Report-FINAL.pdf)

[2] The Advanced Technology Vehicles Manufacturing (ATVM) Loan Program was authorized under section 136 of the Energy Independence and Security Act of 2007. It makes available $25 billion to provide debt capital to the U.S. automotive industry for projects that help vehicles manufactured in the U.S. meet higher millage requirements and lessen U.S. dependence on foreign oil.

[3] Data from Michigan Economic Development Corp.

[4] See the summary of presentation by James M. Forcier of A123 Systems in the next chapter.

[5] This comment proved to be prescient. A123 has since announced bankruptcy and was acquired by Johnson Controls. Johnson has plans to keep the Michigan based production facilities and

the U.S. has the supply base and skilled workforce to sustain a globally competitive industry. These issues present important inter-related questions about the need to stimulate consumer demand, the prioritization of research funding to advance battery technologies, and the need for complementary infrastructure to support the electrification of transportation in the United States.

NATIONAL ACADEMIES SYMPOSIUM

To better understand the progress, challenges, and opportunities facing America's advanced battery industry for electric-drive vehicles, the National Academies' Board on Science, Technology, and Economic Policy (STEP) convened a symposium in Livonia, Michigan, on July 26 and 27, 2010. Organized in cooperation with the Michigan Economic Development Corporation (MEDC) and the Department of Energy, the conference drew leading authorities from government, industry, the U.S. military, academia, and research institutes.

Box A
Competitiveness and Government-Industry Collaboration

In his keynote address, U.S. Senator Carl Levin of Michigan noted that attitudes toward collaboration between government and industry have shifted dramatically in Washington. "A few years ago, anyone who suggested that government work closely with industry was accused of supporting an 'industrial policy.' If that industrial policy label stuck to anything, it was a kiss of death," he recalled.

Now, Senator Levin said, policymakers understand U.S. companies are at a competitive disadvantage because they are competing not just with other companies, but also with other governments that support their domestic industries. These days, "the question no longer is about whether government should be teaming up with industry," he said. "The question is about what we need to do, how we do it, and with what timeline."

Senator Levin predicted the electric-vehicle industry would burgeon and "be important to our country, to our national security, and to the national economy." Nevertheless, he acknowledged that "more challenges lay ahead of us than behind." To see this vision through, government and industry must resolve the challenges. "Tell us what you need to get us there," he said, "and I can commit to you that most of my colleagues and I in the Congress will do everything we can to give you the tools and support you need."

---

workforce that A123 developed, and to incorporate A123 technology into their product lines. http://www.sfgate.com/business/bloomberg/article/A123-Filing-Shows-Struggle-Extending-MIT-Smarts-3971023.php.

In her introductory remarks at the symposium, Dr. Mary Good, of the National Academies STEP Board noted that the conference would inform the Department of Energy and other federal agencies, Congress, and states on the government-industry collaboration required to support the expansion of the market for electric-drive vehicles and "hasten the widespread use of advanced batteries."

A. STRATEGIC IMPORTANCE OF ADVANCED BATTERY MANUFACTURING

Many nations regard the advanced-battery industry as strategic, both as a means of reducing energy use and as an important manufacturing industry. This is no less the case for the United States. Currently, the transportation sector accounts for two-thirds of U.S. petroleum consumption, and two-thirds of that is burned by the 240 million vehicles on U.S. roads.[6] As core components in electricity-powered vehicles, advanced batteries are seen as an important tool to cut U.S. greenhouse gas emissions and limit dependence on imported oil. As speakers at the symposium noted, leadership in the development and manufacture of advanced batteries in the United States is important for the future of the U.S. automobile industry. (See Box B) Despite major U.S. advances in battery research and technology, the United States does not at present lead in the manufacture of this strategic technology.

Box B
Advanced Batteries and the Future of the U.S. Auto Industry: Trading Oil Dependency for Battery Dependency?

Eric Shreffler of the Michigan Economic Development Corporation asserted at the symposium that battery cells and packs are the "the new power train" of future automobiles.[7] Reliance on foreign battery technology and products could thus put the competitiveness of the U.S. auto industry at risk.

In her keynote remarks at the symposium, U.S. Senator Debbie Stabenow (D-MI) said that the last thing the U.S. needs "is to go from a dependence on foreign oil to a dependence on foreign technology. Building the next generation of energy-efficient vehicles is do-or-die for all of the automakers, for the state of Michigan, and for America."[8]

---

[6] The remainder is used by air, rail, and marine and off-road transportation. U.S. Department of Energy data cited in presentation by Patrick Davis.

[7] See the summary of the presentation by Eric Shreffler of the Michigan Economic Development Corp. in the next chapter.

[8] See the summary of the presentation by U.S. Sen. Debbie Stabenow in the next chapter.

U.S. Currently Produces Only About 1 Percent of Lithium-ion Batteries

While American researchers have long been at the forefront of lithium-ion technology, U.S. industry has not dominated the global market for advanced batteries. The industry has been dominated by Asian manufacturers ever since Sony Corporation of Japan marketed lithium-ion batteries for consumer electronics products in 1991. As Mohamed Alamgir of Compact Power noted in his symposium remarks, over this period, a number of U.S. initiatives to manufacture lithium-ion batteries failed, including those by Duracell, Polystor, Motorola, MoliCell, Electro Energy, and Firefly.[9] The U.S. currently produces only about 1 percent of lithium-ion batteries. Japan accounts for 46 percent, South Korea for 27 percent, and China for 25 percent.[10]

Competing in the Market for Advanced Vehicle Batteries

As Ann Marie Sastry of the University of Michigan pointed out at the symposium, battery cells using lithium-ion technology are regarded as the most likely candidates to replace nickel-metal hydride as the most common source of power storage in electric vehicles.[11] A lithium ion battery produces electrical charges by lithium ions that flow between an anode plate and a cathode plate. The liquid chemical mixture inside the battery, known as electrolyte, contains lithium salts and an organic compound. Pike Research predicts the market for lithium-ion batteries for transportation will grow over 700 percent, from $2.0 billion annually in 2011 to greater than $14.6 billion by 2017.[12]

The more demanding requirements of lithium-ion batteries for cars rather than consumer electronics present an opportunity for the U.S. to become an important player in the industry. Although U.S. start-ups and national laboratories continue to be leading sources of innovation in the lithium-ion battery "chemistries," or the coatings and materials used in the cathode and

---

[9] According to analysis by Ralph Brodd, "The U.S. battery companies "opted out" of volume manufacturing of Li-ion batteries, primarily because of a low return on investment compared with their existing business, the significant time and investment required from conception to commercialization, and the time and expense required to establish a sales organization in Japan to access product design opportunities and take advantage of them." See Ralph J. Brodd, "Factors Affecting U.S. Production Decisions: Why Are There No Volume Lithium-Ion Battery Manufacturers in the United States." Gaithersburg MD: NIST GCR 06-903, December 2006. Access at http://www.atp.nist.gov/eao/gcr06-903.pdf. Compact Power, which is backed by LG of South Korea announced in late 2012 that they are furloughing workers at their production facility in Michigan. Compact Power is contracted to provide batteries for the Volt and the Ford Focus, but to date they have not produced batteries at their Michigan plant, having satisfied current demand with batteries manufactured in Korea. http://www.theblaze.com/stories/how-many-chevy-volt-batteries-will-150-million-make-hint-less-than-one/

[10] See the summary of the presentation by Patrick Davis of Department of Energy in the next chapter.

[11] For an example of such analysis, see Rod Loach, Dan Galves, Patrick Nolan, "Electric Cars: Plugged In. Batteries Must be Included," Deutsche Bank Securities Inc., June 9, 2008.

[12] Pike Pulse Report: Electric Vehicle Batteries, February 2012,

anode, some analysts have expressed the concern that U.S. industry will not be able to compete successfully in the market for advanced vehicle batteries.

Currently, the U.S. remains far behind its competitors in Asia in high-volume manufacturing capability. Japan has targeted lithium-ion batteries for vehicles since 1992, when the Agency of Industrial Science and Technology and the Ministry of International Trade and Industry established the New Sunshine Program.[13] South Korea's government has committed $12.5 billion in a bid to become the world's leading producer of advanced batteries.[14] China, which is gaining fast, heavily subsidizes domestic battery manufacturers and requires foreign battery companies to manufacture in China if they wish to sell there.[15]

The Demand for Electrified Vehicles

Moreover, demand for electrified vehicles has been stronger outside of the United States. Higher fuel prices, in large part due to high taxes, make hybrids and plug-ins a more economically attractive option in Europe. Other nations have acted more to develop their domestic market for electrified vehicles by offering subsidies and installing battery-charging infrastructure. China, for instance, awards $8,800 to domestic automakers for every electric vehicle sold. Some Chinese regional governments offer additional subsidies.[16] Thanks largely to such policies, Pike Research predicts Asia will account for 53 percent of global demand for electrified vehicles in 2015—more than the U.S. and Europe combined.[17]

Currently, demand for electrified vehicles is being held back by the high cost of a typical hybrid battery pack.[18] Although price has dropped by more

---

[13] See Alamgir presentation. Japan's New Sunshine Program established a 10-year research program for lithium-ion batteries that set very ambitious targets for the time for power output, battery density, and cycle life. See Rikio Ishikawa, "Current Status of Lithium-Ion Production in Japan," Central Research Institute of Electric Power Industry, Tokyo (http://www.cheric.org/PDF/Symposium/S-J3-0003.pdf).

[14] Yonhap News Agency, "S. Korea Aims to Become Dominant Producer of Rechargeable Batteries by 2020," July 11, 2010.

[15] Forcier presentation, op. cit. For a review of Chinese policies to promote the Chinese automotive industry. See, Terrence Stewart, et al. "China's Support Programs for Automobiles and Auto Parts under the 12th Five-Year Plan." Washington, DC: Law Offices of Stewart and Stewart, 2012. The report notes that certain policies have been found to violate commitments made by China on joining the WTO. Access at http://www.stewartlaw.com/stewartandstewart/Portals/1/Douments/S%20&%20S%20China%20Auto%20Parts%20Subsidies%20Report.pdf. For a review of the impact of Chinese state capitalism on U.S. innovation, see Andrew Szamosszegi and Cole Kyle, "An Analysis of State-owned Enterprises and State Capitalism in China", Washington, DC: U.S.-China Economic and Security Review Commission, October 26, 2011. For a review of national support around the world for emerging industries including advanced batteries, see National Research Council, *Rising to the Challenge, U.S. Innovation Policy in the Global Economy*, C. Wessner and A. Wm. Wolff, eds., Washington, DC: 2012, Chapter 6.

[16] Forcier presentation, op. cit

[17] Forcier presentation, op. cit.

[18] Data are for batteries discharging 25 kilowatts of power.

than two-thirds since 1997, and while densities and life cycles have more than doubled, the battery back for plug-in hybrid cars still costs around $2,500.[19] Unless gas prices skyrocket, some analysts believe costs must drop by around two-thirds and that battery size must shrink dramatically before most consumers see the payoff of abandoning gas-powered cars and paying a $6,000 to $12,000 premium for a battery-powered car.

The resulting slow pace of adoption of Electric Drive Vehicles is making it difficult for U.S. Battery Companies to survive and a domestic supply chain to develop.[20] The emergence of the US battery industry therefore is likely to depend on markets other than electric vehicles such as Consumer Electronics and Grid Storage. Established companies with good balance sheets and a perspective on long-term investment will be necessary.

B. FEDERAL INITIATIVES TO ESTABLISH A U.S. ADVANCED BATTERY INDUSTRY

Symposium participants noted that the U.S. government has recently taken a number of active steps to establish a strong U.S. advanced battery industry and market for electrified vehicles.[21]

- The Department of Energy's Vehicle Technologies Program has made lithium-ion battery research and development a high priority since 2000.[22]
- The Department of Energy also leads a government-industry partnership called the U.S. Advanced Battery Consortium, which funds projects aimed at commercializing new battery technologies and sets cost and performance targets for the industry.[23]

---

[19] Data cited by David Howell of the Department of Energy in his presentation, which is summarized in the next chapter.

[20] Enerl is now in Chapter 11 Bankruptcy. See Businessweek, "Ener1, Battery Maker, Seeks Chapter 11 Bankruptcy Protection," February 08, 2012. Short on cash, A123 Systems had signed a non-binding memorandum with Wanxiang Group Corporation, a Chinese largest auto parts manufacturer, seeking additional investments of up to $450 million. As one analyst has put it, "this investment for Wanxiang is almost certainly about acquiring A123's technology and business contacts at a discount…" See Tom Konrad, "A123's Deal with China's Wanxiang Would Value the Stock at $0.55 a share." altenergystocks.com, August 19, 2012.

[21] On March 6, 2012, President Obama announced a $4.7 billion proposal to expand electric vehicles. The EV-Everywhere Challenge is focused on advancing electric car technologies while reducing costs. The EV-Everywhere Challenge is the second of the Energy Department's Grand Challenges, following the model of the $1/watt SunShot Challenge, which seeks to make solar power directly cost-competitive with electricity from fossil fuels by the end of the decade. On March 9, 2012, President Obama called for a $1 billion "National Network for Manufacturing Innovation," that will help develop up to 15 manufacturing "Institutes" to foster innovation around the country.

[22] The Vehicle Technologies Program is administered by the Energy Efficiency and Renewable Energy Office of the Department of Energy. It funds projects aimed at developing "leap frog" technologies that will lead to more energy-efficient and environmentally friendly transportation. See presentation by David Howell of the Department of Energy's Vehicle Technologies Program.

[23] The United States Advanced Battery Consortium is a collaboration between the Department of Energy and the United States Council for Automotive Research, whose members consist of General

- The 2009 Recovery Act grants to battery cell, pack, and materials companies are also expected to boost U.S. manufacturing capacity to 1 million batteries a year by 2015.[24]
- The battery industry will also benefit from complementary investments in the smart-grid, funded by $4.5 billion in Recovery Act funds.

Additional Federal Initiatives

Symposium participants noted that the federal government also supports vehicle electrification in other ways:

- **Funding for Research and Commercialization**
    - The Advanced Research Projects Agency-Energy (ARPA-E), a new Department of Energy program that funds "transformational" energy-technology R&D, has funded $100 million for energy-storage research.[25]
    - Battery manufacturers are expected to share some of the $25 billion set aside under the government's Advanced Technology Vehicle Manufacturing Program to speed the commercialization of advanced battery technology.[26]
- **Tax Incentives and Credits**
    - The Advanced Energy Manufacturing Tax Credit program provides $2.3 million to companies to cover 30 percent of investments in new, expanded, or refurbished manufacturing plants producing renewable-energy equipment.[27]
    - U.S. consumers buying electrified vehicles also can receive tax deductions.
    - The U.S. government also has recently begun offering loan guarantees to green-technology projects, tax credits for renewable energy "property," and greater access to export financing.[28]
    - Congress also has been expanding incentive programs to include suppliers and light trucks.[29]

Motors, Ford, and Chrysler. The group's stated mission is "to develop electrochemical energy storage technologies that support commercialization of fuel cell, hybrid, and electric vehicles."

[24] See the summary of the presentation by Patrick Davis of the Department of Energy in the next chapter.

[25] See the summary of the presentation by David Howell of the Department of Energy in the next chapter.

[26] See the summary of the presentation by Patrick Davis of the Department of Energy in the next chapter

[27] [27] See the summary of the presentation by Sen. Stabenow. The Advanced Energy Manufacturing Tax Credit was authorized in Section 1302 of the American Recovery and Reinvestment Act and also is known as Section 48C of the Internal Revenue Code. It authorizes the Department of Treasury to award $2.3 billion in tax credits to cover 30 percent of "investments in advanced energy projects, to support new, expanded, or re-equipped domestic manufacturing facilities."

[28] See the summary of the presentation by Michael Reed in the next chapter.

[29] See the summary of the presentation by Sen. Stabenow.

- **Standards**

Tougher federal and state environmental standards are being proposed to boost the industry. The Obama Administration wants to set a target of reducing greenhouse gas emissions by at least 30 percent by 2016.[30] California has even more aggressive emission targets. The state is raising requirements on automakers to sell a certain number of zero-emission vehicles and wants the carbon-intensity of all fuels cut by 10 percent.[31]

- **Procurement**

The U.S. military is another important driver of advanced batteries.[32] The U.S. Army, which has one of the world's largest vehicle fleets, has committed to cutting its fuel consumption by 20 percent in the next 10 to 15 years. At the same time, new weapons systems and other requirements are boosting the need for power in combat and non-combat vehicles.[33] The logistical challenges of transporting fuel into the battlefield present another strong motive for reducing fuel use. Through the Tank-Automotive Command Research, Development, and Engineering Center (TARDEC), which is based in the Detroit area, and the Army Research Laboratory, the Army collaborates with the Department of Energy and industry on research and development in batteries, new materials, and electrical systems.[34]

Getting in the Game

Despite entering the industry late, a number of speakers maintained that the U.S. still has an opportunity to become a major global player in advanced batteries. One reason is that the industry is still young. Most analysts predict that electrified cars will account for only 2 percent to 3 percent of the U.S. market in

---

[30] The U.S. Environmental Protection Agency and the Department of Transportation's National Highway Traffic Safety Administration (NHTSA) are finalizing greenhouse gas-emission standards for model years 2012 to 2016 under the Energy Policy and Conservation Act. For details, see http://www.epa.gov/oms/climate/regulations/420f10014.htm.

[31] See the summary of the presentation by Daniel Sperling of the University of California at Davis in the next chapter. For an international comparison of vehicle emission targets, see Feng An, et al. Global Overview on Fuel Efficiency and Motor Vehicle Emission Standards: Policy Options and Perspectives for International Cooperation." New York: United Nations Commission on Sustainable Development, CSD19/2011/BP3, May 2011. See in particular, Figure 5 on page 18. Access at http://www.un.org/esa/dsd/resources/res_pdfs/csd-19/Background-paper3-transport.pdf.

[32] In a February 29, 2012 speech at the Energy Innovation Summit of the Department of Energy (ARPA-E), Deputy Defense Secretary Ashton Carter told the audience the Pentagon could be an early adopter of innovations and push the technological edge out further than other entities because it is willing to pay more for better capabilities. It could also buy new hardware in vast quantities, further driving technological refinements that would reduce costs, Carter said. Those lower prices might then lead to wider adoption of such new technologies.

[33] See the summary of the presentations by Grace Bochenek and Sonya Zanardelli of the U.S. Army Tank and Automotive Research, Development, and Engineering Center in the next chapter.

[34] See the summary of the presentation by Sridhar Kota in the next chapter.

2015 and 5 percent in 2020.[35] Many industry experts also believe lithium-ion batteries will have to evolve through several more generations of technology and manufacturing improvements before they are affordable, efficient, and light enough to win wide consumer acceptance for electric cars.

C. MICHIGAN SEIZES THE INITIATIVE

Michigan began studying ways of capturing the electric-vehicle and advanced-battery industries in 2005, well before the federal government got involved.[36] As Greg Main, CEO of MEDC, the state's economic development agency, noted in his symposium presentation, this sector was recognized as an opportunity to diversify Michigan's manufacturing base into clean-energy products. Officials believed Michigan's strong base in automotive manufacturing and engineering provided a clear advantage in the nascent industry of lithium-ion batteries for cars.

Michigan's decision to offer generous incentives to battery manufacturers "sent a clear signal that Michigan is very serious about being a leader in this industry," Michigan Governor Jennifer Granholm said in her address.

Those early corporate commitments paid off when the Department of Energy awarded $1.3 billion of the $2.4 billion allocated for advanced-battery manufacturing projects under the American Recovery and Reinvestment Act of 2009 to Michigan-based factories, including battery plants by A123, Johnson Controls-Saft, Dow Kokam, and Compact Power, a unit of South Korea's LG Chem.[37] In her remarks, Governor Granholm noted that this investment has helped to leverage nearly $6 billion in private investment in the 16 advanced battery and battery technology projects underway in Michigan.

---

[35] Pike Research predicts the penetration rate of hybrid and plug-in vehicles will be 2.41 percent in 2015.

[36] With support from New York State, General Electric announced in 2009 the building of a $100 million battery manufacturing facility in the Albany region GE has also invested $70 million in A123 Systems with which it has partnered to finesse battery management, battery safety, and fusing systems. Researchers from GE are also working on a dual-battery system with the Department of Energy. See Cora Nucci, "GE to build advanced battery plant in NY state." Information Week, May 12, 2009.

[37] See the summary of the presentation by Greg Main in the next chapter. Dow Kokam will complete its $322 million Midland battery plant in 2012. That plant is supported by a $161 million Energy Department loan and $180 million in tax incentives from the state. Johnson Controls Inc. opened its lithium-ion battery cell plant in July 2011. LG Chem is also building a $300 million factory in Holland, MI to produce batteries for the Chevrolet Volt and electric Ford Focus.

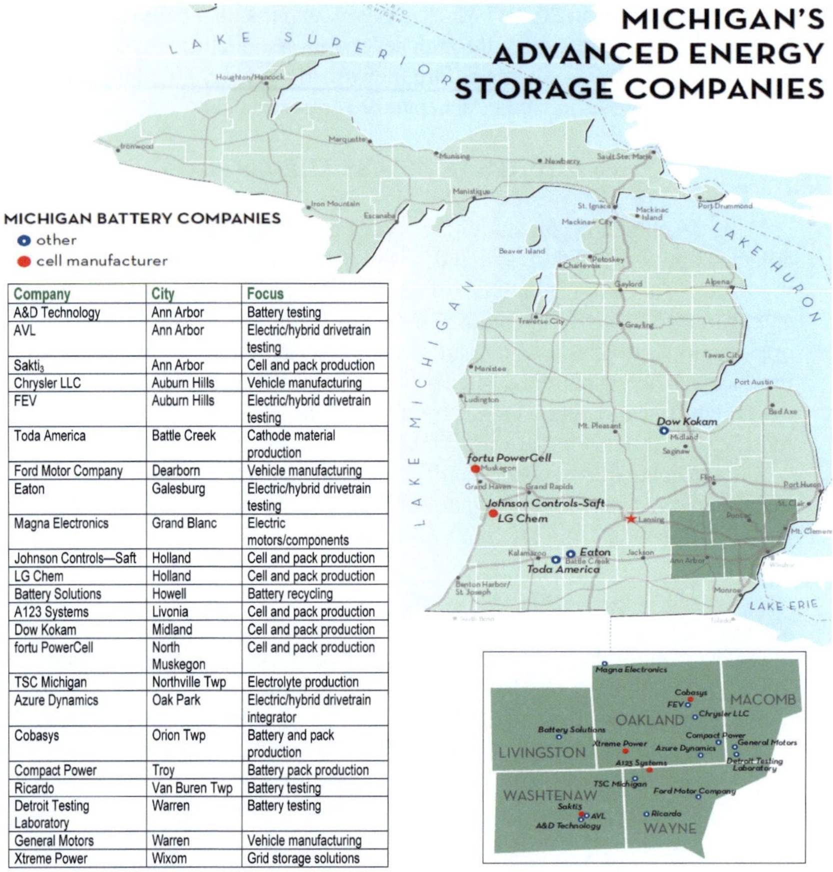

Company	City	Focus
A&D Technology	Ann Arbor	Battery testing
AVL	Ann Arbor	Electric/hybrid drivetrain testing
Sakti$_3$	Ann Arbor	Cell and pack production
Chrysler LLC	Auburn Hills	Vehicle manufacturing
FEV	Auburn Hills	Electric/hybrid drivetrain testing
Toda America	Battle Creek	Cathode material production
Ford Motor Company	Dearborn	Vehicle manufacturing
Eaton	Galesburg	Electric/hybrid drivetrain testing
Magna Electronics	Grand Blanc	Electric motors/components
Johnson Controls—Saft	Holland	Cell and pack production
LG Chem	Holland	Cell and pack production
Battery Solutions	Howell	Battery recycling
A123 Systems	Livonia	Cell and pack production
Dow Kokam	Midland	Cell and pack production
fortu PowerCell	North Muskegon	Cell and pack production
TSC Michigan	Northville Twp	Electrolyte production
Azure Dynamics	Oak Park	Electric/hybrid drivetrain integrator
Cobasys	Orion Twp	Battery and pack production
Compact Power	Troy	Battery pack production
Ricardo	Van Buren Twp	Battery testing
Detroit Testing Laboratory	Warren	Battery testing
General Motors	Warren	Vehicle manufacturing
Xtreme Power	Wixom	Grid storage solutions

FIGURE 1 Michigan's Advanced Energy Storage Companies.
SOURCE: Eric Shreffler, Presentation at July 26-27, 2010 National Academies Symposium on "Building the U.S. Battery Industry for Electric Drive Vehicles: Progress, Challenges, and Opportunities."

Box C
Growing an Advanced Battery Cluster in Michigan

In her symposium address, Michigan Governor Jennifer Granholm predicted that the state "is well on its way to becoming the advanced battery capital of the world. A whole advanced battery supply chain is taking root from the Detroit area to the shores of Lake Michigan."

This optimistic view was echoed by Greg Main of the Michigan Economic Development Corporation: "This is a very exciting time for our country and our state. We are giving birth to an entire new industry in North America."

Box D
Targeting the Heart of the Value Chain

The MEDC began by targeting "the heart of the value chain" for batteries—the cell and battery-pack factories and vehicle electrification programs of major auto makers. "We wanted to solidify and cement as much of that here in Michigan as possible," Mr. Shreffler said. The MEDC saw a need for "very aggressive incentives."

Michigan's Policy Approach

Advanced batteries was one of five promising renewable-energy clusters the MEDC identified, explained Eric Shreffler, who leads the MEDC's advanced energy storage program. Michigan also sought to develop clusters in the technologies related to materials, bio-energy, solar cells and panels, water technology, and wind power. The MEDC formed teams to devise strategies for each cluster.

Besides being a major new growth industry, the MEDC viewed advanced batteries as strategically important because they will be the core technology of future automobiles, Mr. Shreffler said. "Michigan did not want to stand by and cede leadership in power-train development to other states and countries". By being the first state to offer strong incentives, Michigan wanted to "send a signal [that] we are serious about developing this ecosystem in this state" and increase its odds of attracting any potential federal funding, Mr. Shreffler explained.

The MEDC first targeted cell and battery pack manufacturing and vehicle electrification programs. Michigan launched the Centers of Energy Excellence Program, the first program allowing the MEDC to offer grants to for-profit companies, Mr. Shreffler said.[38] It granted $13 million to Sakti3 and A123 on condition they secure federal funds and establish university partnerships.

The other major action was the Michigan Advanced Battery Tax Credits (MABC) program.[39] The response from industry was so strong that the legislature boosted funding from $335 million to $1.02 billion. Of that, $600 million went to six companies committing to build fully integrated cell manufacturing facilities: Johnson Controls-Saft, LG Chem/Compact Power,

---

[38] Michigan's Centers of Energy Excellence Program was established under Senate Bill 1380, Public Act 175. In the program's first phase, the Michigan Strategic Fund Board awarded $43 million in grants in 2008. For-profit companies receiving grants must secure matching federal funds and financial backing. Public Act 144 of 2009 allowed a second phase of the COEE program.

[39] See the summary of remarks by Eric Shreffler. Michigan's Advanced Battery Tax Credits initiative was created through an amendment to the Michigan Business Tax Act, Public Act 36 of 2007, to allow the Michigan Economic Development Corporation to extend tax credits for battery pack engineering and assembly, vehicle engineering, advanced battery technology development, and battery cell manufacturing.

FIGURE 2 Michigan's energy storage industry: supply chain investments.
SOURCE: Eric Shreffler, Presentation at July 26-27, 2010 National Academies Symposium on "Building the U.S. Battery Industry for Electric Drive Vehicles: Progress, Challenges, and Opportunities."
NOTE: The JCI Saft venture dissolved since the date of this presentation.

A123, Dow-Kokam, fortu Powercell, and Xtreme Power. Michigan refunds up to $100 million of their capital investment, Mr. Shreffler explained. Another $225 million went to battery pack manufacturers, who receive a credit for each pack they assemble in Michigan. The $1.3 billion in grants through the Recovery Act mainly went to these same companies.

Michigan's pipeline of new projects "continues to be very full," Mr. Shreffler said. They include a cathode materials plant by Toda America, battery testing facilities by AVL and A&D Technology, electric motor components by Magna, energy-storage solutions by Xtreme Power, and electric drive-train testing by Eaton.

D. REGAINING U.S. LEADERSHIP IN BATTERY TECHNOLOGY

Investing in the 'Manufacturing Commons'

Although "the stars are all aligned" now for the U.S. to regain global leadership in battery technology, there are currently not a sufficient number of battery and electric vehicle assembly plants to make the U.S. global competitive,

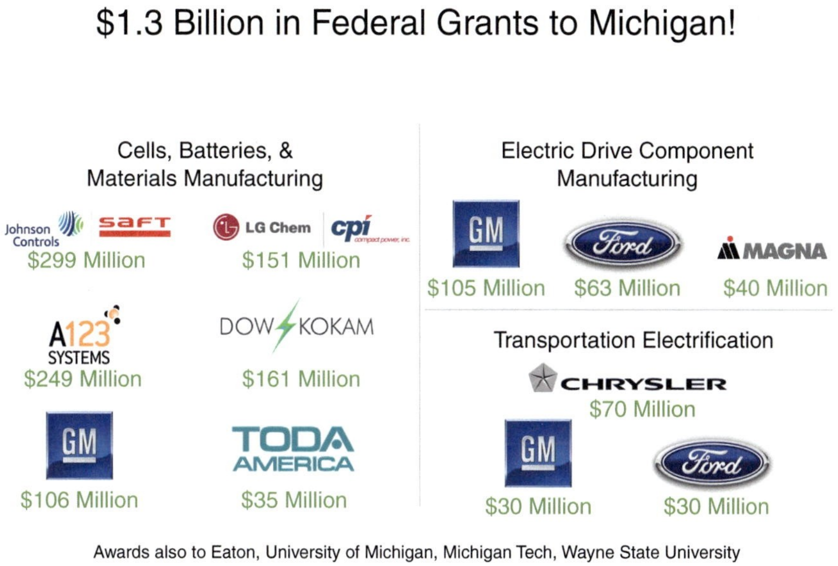

FIGURE 3 Michigan's energy storage industry: federal grants.
SOURCE: Eric Shreffler, Presentation at July 26-27, 2010 National Academies Symposium on "Building the U.S. Battery Industry for Electric Drive Vehicles: Progress, Challenges, and Opportunities."

Sridhar Kota of the White House Office of Science and Technology said in his opening remarks at the symposium. America also must invest in basic research and what he referred to as the "manufacturing commons," which is a combination of elements that together make up an ecosystem that is conducive to manufacturing.[40]

This commons, which is needed to support large-scale production, includes engineering R&D, management expertise, a skilled workforce, access to capital, a components industry, production equipment, industry standards, and product platforms. Some of these capabilities have eroded due to cutbacks in corporate research and decades of offshore outsourcing, Dr. Kota said. "If you don't have those manufacturing commons in place, we are not going to be able to make next-generation products," he said. The Administration outlined its strategy for revitalizing American manufacturing in a white paper released in December 2009.[41]

---

[40] For an analysis of the importance of the "industrial commons," see Gary P. Pisano and Willie C. Shih, "Restoring American Competitiveness," *Harvard Business Review*, July 2009.
[41] See "A Framework for Revitalizing American Manufacturing," Executive Office of the President, Dec. 16, 2009 (http://www.manufacturing.gov/pdf/20091216-manufacturing-framework-final_embargoed.pdf).

The federal government recently has boosted efforts to develop a manufacturing commons for advanced batteries. In addition to Recovery Act funds for advanced-battery and smart-grid projects, Dr. Kota observed that the Obama Administration has substantially increased the advanced manufacturing tax credit program from $2 billion to $7 billion. Other incentives include the Department of Energy's 1703 and 1705 loan guarantee programs[42] and the 1603 program that gives cash grants in lieu of tax credits for renewable-energy projects.[43] In the battery industry, such programs have complemented aggressive incentives offered by states such as Michigan

At the basic research level, Dr. Kota noted that the Obama Administration has made advanced vehicle technologies one of its six top priorities for research funding.[44] The Department of Energy's ARPA-E program is working on new composites for vehicles and "potential breakthroughs in new battery chemistries that are two or three or five times better than current technologies," he said, and manufacturing technologies "that could change the game altogether."

In applied research, Dr. Kota reported that President Obama's FY 2011 budget calls for investing $12 million for university innovation centers that focus on developing proofs of concept and prototypes and an additional $10 million for nano-manufacturing. The Department of Commerce and the Office and Science and Technology Policy have R&D commercialization programs at universities. The federal government funds programs around the U.S. to train the advanced-manufacturing workforce and expand the pool of engineers.

Further Help from Congress

Speaking at the symposium, U.S. Senator Stabenow noted that efforts are also underway in Congress to increase federal help for advanced vehicle technologies. "It is incredibly important that we ramp this us as fast as we can." Senator Stabenow is a member of the Senate Finance, Energy, and Agriculture committees and has co-authored legislation including the Cash for Clunkers Program and the Advanced Energy Manufacturing Tax Credit.

---

[42] Section 1703 of Title XVII of the Energy Policy Act of 2005 ("EP Act 2005") authorizes the Department of Energy to issue loan guarantees to acceleration commercialization of technologies that "avoid, reduce, or sequester air pollutants or anthropogenic emission of greenhouse gases." Section 1705 of the EP Act is a temporary program set up under the American Recovery and Reinvestment Act authorizing the Department of Energy to make loan guarantees to renewable energy systems, electric transmission systems and leading-edge bio-fuels projects that commence construction no later than September 30, 2011.

[43] Section 1603 of the American Recovery and Reinvestment Act created a program administered by the U.S. Department of Treasury that extends grants covering between 10 percent and 30 percent of the cost of certain renewable-energy property.

[44] From M-1-30 Memorandum for the Heads of Executive Departments and Agencies, by Peter R. Orszag, director of Office and Management and Budget, and John P Holdren, director of Office of Science Technology Policy, "Science and Technology Priorities for FY 2012 Budget," Executive Office of the President, July 21, 2010.

Senator Stabenow noted that Congress is looking to expand the Advanced Technology Vehicle Manufacturing program beyond car and battery manufacturers. She further noted that she is co-sponsoring legislation with Representative Gary Owens (D-MI) are to extend help to medium- and heavy-duty trucks. Federal loan program for factory retooling, meanwhile, has been amended to include medium- and heavy-duty vehicle suppliers.

Infrastructure: To address infrastructure, Senator Stabenow said that a bipartisan bill that she has sponsored calls for the Department of Energy to help 15 U.S. communities develop charging stations for hybrids and plug-ins and help consumers get what they need to charge cars at home.[45] "We want to create models of how to develop that infrastructure as quickly as possible," she said.

Tax Credits: To boost demand or electrified vehicles, Senator Stabenow said that she and other legislators are seeking to expand the current $7,500 tax credit now given to purchases of plug-in hybrid cars. This program applies only to the first 2,500 purchases and expires in 2014. She said that she is working on legislation to have these credits awarded at the time a car is being purchased at a showroom, rather than as a tax deduction the following year, noting that like the "Cash for Clunkers" program, an upfront rebate has a bigger impact on spurring demand. She also proposed that such credits apply to commercial trucks. She also noted another Senate bill focusing on generating demand encourages federal agencies to purchase electrified vehicles.

Trade Policy: Fair trade is another priority, Senator Stabenow said. In response to Chinese policies directing the Chinese government do business only with Chinese companies,[46] she said that she has co-sponsored a bill that would bar U.S. government purchases of Chinese products until Beijing signs a World Trade Organization agreement on government procurement.[47] "We have to have access to markets if we are going to meet our exporting goals."

The DoE's Vehicle Technology Strategy

According to Patrick B. Davis the program director of DoE's Energy Efficiency and Renewable Energy Vehicles Technology Program, the funds

---

[45] The Promoting Electric Vehicles Act of 2010 (S. 3495) sponsored by Sen. Byron Dorgan (D-ND), Sen. Debbie Stabenow (D-MI), and Sen. Lamar Alexander (R-TN) calls for providing incentive programs to create "deployment communities" across the U.S. stations for purchasing electric vehicles and set up charging facilities. The Senate Energy and Natural Resources Committee approved the bill on July 27, 2010.

[46] China's 15-year plan for science and technology says the government should practice a "first-buy policy for major domestically made high-tech equipment and products that possess proprietary intellectual property rights." See Sec VIII, 3 of "The National Medium- and Long-Term Program for Science and Technology Development (2006-2020): An Outline," pg. 54, State Council of China.

[47] The China Fair Trade Act of 2010 (S. 3505) was introduced on June 17, 2010, by Sen. Lindsay Graham (R-S.C.), Sen. Debbie Stabenow (D-MI), Sen. Russ Feingold (D-MN), and Sen. Sherrod Brown (D-OH). It would bar the U.S. government from purchasing Chinese products until China agrees to the Agreement on Government Procurement of the World Trade Organization.

made available through the Recovery Act present a "once-in-a-lifetime opportunity" to establish a U.S. advanced-battery industry.

Funding: Although most of the 48 battery-related projects funded through the Recovery Act involve cell and battery manufacturing, Mr. Davis noted that DoE's strategy is to establish the entire supply chain. Accordingly, the Department of Energy has awarded funds to producers of lithium, electrolytes, separators, and materials for cathodes and anodes. It also funded lithium recycling projects. To fund this technology beyond the Recovery Act, DoE's Vehicle Technologies Program budget is set to grow to $121 million by 2011. The DoE's Office of Science, ARPA-E program, and Office of Electricity also are active in developing innovative battery technologies.

Deployment: The Department of Energy also funds projects to demonstrate and deploy innovative electric vehicles and charging infrastructure. So far, Mr. Davis reported that eight grants have been awarded to projects that will deploy 10,000 electric-drive vehicles, ranging from light-duty trucks to passenger busses, as well as home and public-access chargers across the nation. The DoE's Clean Cities program, meanwhile, works with 86 coalitions in 45 states to introduce thousands of hybrid and electric vehicles and charging stations.

Targets: Mr. Davis also noted that the Department of Energy has set ambitious targets to lower battery costs and boost performance. Current lithium-ion batteries for cars cost an average of $800 per kilowatt-hour in a laboratory setting. The goal, he said, is to cut that to $500 per kilowatt-hour in 2012 and $300 in 2014 for a plug-in hybrid. The Department of Energy also wants drastic cuts in greenhouse gas emissions, to around 50 grams of CO_2 equivalent per mile compared to an average of 430 grams now with conventional cars and some hybrids. That probably will not occur for several more decades, he predicted, when cars run entirely on electricity. Electric-drive technology, therefore, "is very important."

E. THE MILITARY'S ELECTRIFICATION DRIVE

The U.S. military is another important promoter of advanced vehicle technologies, explained John Pellegrino of the Army Research Laboratory and Grace Bochenek of TARDEC in their presentations. TARDEC oversees maintenance of the Army's 400,000-vehicle fleet and development of "next-generation capabilities," Dr. Bochenek explained.

At the same time, it is trying to slash energy use. Dr. Pellegrino and Dr. Bochenek explained that new weapons systems and other requirements are boosting the power needs of Army vehicles. Concurrently, for logistical reasons, the Army also wants combat vehicles to run longer without refueling and to cut the need for trucking convoys to haul fuel.[48]

[48] According to Secretary of the Navy Ray Mabus, when you factor in all the costs of transporting fuel by truck or air to a forward base in Afghanistan — that is, guarding it and delivering it over

The Army has ambitious plans to introduce electrified vehicles into its fleet and develop lighter-weight, higher-density batteries. It also requires advanced batteries for soldiers, mobile devices, and unmanned aerial vehicles. According to Dr. Pellegrino, advanced batteries are an important element in each scenario for reducing energy use. What's more, he noted, the Army's needs differ from those of the commercial sector because "we see more extreme environments than the average citizen." Reliability is vital, and safety is extremely important because equipment can come under fire.

There are major opportunities for partnerships with the private sector. Over the past five or 10 years, the Army has been "doing much, much more early collaboration with industry," Dr. Pellegrino said. "We don't want each of those vehicles to cost $1 billion. It is only by leveraging and working with the commercial market" that the high production volumes can be attained that will reduce costs.

The U.S. Army's Special Needs

One top Army objective is to achieve greater "energy independence" for tactical units so that soldiers and vehicles can operate days or weeks longer without refueling. In Kuwait, the Army moves around 431 million gallons of fuel a year. That translates into 140,000 trucks, 9,300 convoys, and 644,000 trips by soldiers each year. Cutting fuel use by just 1 percent "reduces the number of soldiers you have to put in harm's way by 6,444, which is significant," Dr. Bochenek noted. The Army also wants to sharply boost the fuel-efficiency of future light tactical vehicles to 61 ton-miles per gallon, a nearly 50 percent improvement from current Humvees. It also wants tanks that can operate two or three days without refueling and Stryker armor cars with cruising ranges of up to 360 miles, she said.

Dramatic improvements in batteries are required to meet the ever-rising power requirements of combat vehicles. In World War II, the Army consumed about one gallon of gas a day per soldier, Dr. Bochenek said. Today, it consumes 20 gallons. Half is used to generate electricity for jammers, satellite remote sensing equipment, systems for defeating improvised explosive devices, and active protection systems.

The needs for lighter, more powerful batteries will only grow. A high Army priority is to fit combat vehicles with Silent Watch capability, for example, enabling them to operate essential systems while stationary without running the engine. Future light tactical vehicles will require 40 kilowatts of power, compared to 10 kilowatts now, Dr. Bochenek said. Future ground combat systems will need nearly 50 kilowatts.

Cost reduction also is critical. Although lithium-ion battery packs for light tactical vehicles weigh one-third as much as advanced lead-acid batteries

---

mountains — a single gallon of gasoline "could cost up to $400" once it finally arrives. See *The New York Times*, The U.S.S. Prius, Thomas L. Friedman, December 18, 2010

and produce 50 percent more power, they cost nearly 20 times as much—around $10,000 each. To achieve each of these goals, Dr. Bochenek said, "We really need to increase the density and at the same time reduce the weight and volume" of batteries.

F. PRIVATE SECTOR STRATEGIES

Ford's Diversification Strategy

Forecasting the scale and nature of future demand is a major challenge for passenger car makers. In the U.S., some 70 percent of electric cars sold in 2020 are projected to be hybrids[49] and another 25 percent plug-ins,[50] noted Nancy Gioia, Ford Motor's director of global electrification. In Europe by contrast, plug-ins and all-battery electrics[51] are expected to account for more than half of the market.[52]

Still, Ford keeps investing in vehicles powered by conventional internal combustion engines because they are expected to dominate the market for decades. Electrified cars accounted for only 1 percent of Ford's sales in 2010, Ms. Gioia said. It aims to boost that to 2 percent to 5 percent in five years and up to 25 percent in 2025.

Ford's strategy is to offer a full portfolio of electrified cars and small trucks. It is marketing hybrid versions of its Fusion sedans and Escape cars, for example, and in 2010 launched the Transit Connect line of small commercial vehicles. The Fusion Electric small car was introduced in 2011.

Electric vehicles, however, are no "silver bullet" to assure a sustainable business, Ms. Gioia said. "We will see growth in electrification," she said. "But we also are going to technologies that continue to improve [the efficiency of] petrol and diesel solutions..." Improvements in current technology will make a faster and greater impact on national fuel consumption because they don't require new transportation infrastructure, she said.

Battery costs are the "Achilles heel" of electrified cars, Ms. Gioia said. "We need to go through two to three cycles of innovation and then scale up appropriately" to have a product affordable to most customers, she said. Ford wants battery suppliers to cut the hybrid pack costs from a projected $750 per kilowatt hour in 2012 to $250 in 2020, she said.

---

[49] A hybrid car has a dual mechanical and electric power train. It operates on battery power for limited times, such as while starting the engine, during acceleration, or driving for short distances. After that, the internal combustion engine takes over.

[50] A plug-in hybrid car has a battery that can be recharged overnight from an electrical socket and store enough electricity to drive a car for certain distance, typically 10 to 100 miles.

[51] In an all-battery electric car, 100 percent of propulsion comes from electric motors energized by power stored in the battery. They do not have dual mechanical and electric power trains.

[52] Data compiled by Ford Motor from studies by JP Morgan, Credit Suisse, Boston Consulting Group, A. T. Kearney, and Roland Berger.

Temperature control, energy density, and the number of real-world charge and discharge cycles also remain serious challenges. The current battery for the Focus all-battery electric car produces 23 kilowatt hours, adds 500 pounds to the vehicle, and is 125 liters in size. "That is whomping big to fit into a car," Ms. Gioia said. "Not until third-generation batteries (weighing around 250 pounds and arriving in an additional five or six) will batteries truly be replaceable in cars. If it turns out customers really want electric cars with a 200-mile range, rather than 100, she added, "that just exacerbates this challenge."

LG: The Importance of Deep Pockets

Federal financial help has been vital for the fledgling U.S. battery industry. In his presentation, Compact Power Research Director Mohamed Alamgir cited numerous American battery companies—including three he worked for—that either went out of business or abandoned lithium-ion in the 1990s for lack of funding and because they could not compete with better-financed Japanese competitors. Colorado-based Compact Power, established in 2000, received its initial funding from the Department of Energy and "was kept alive" through lean times from 2003 to 2006 by funds from the Department of Energy and the U.S. Advanced Battery Consortium, he said.

Now Compact Power, a unit of South Korea's LG Chem, is building a large plant in Holland, Mich. The Department of Energy and LG Chem each are contributing $151 million to the complex, which will start manufacturing lithium-ion cells in 2012 and eventually make electrodes. The plant will be capable of making up to 20 million cells a year, enough for more than 50,000 vehicles, and employ 300 people.

Being part of LG Chem offers several advantages, Mr. Alamgir said. The Korean petrochemical giant is the world's third-largest producer of rechargeable lithium-ion batteries, mainly for consumer devices such as notebook computers and mobile phones, as well as lithium-ion cells to Ford and GM. It also is part of the $113 billion LG Group. Having deep pockets is important "to survive in this industry," Mr. Alamgir said.

LG Chem is a vertically integrated company. It designs and manufactures battery packs and electrical management systems and develops power and signal architectures, thermal management solutions, and test and validation services. Most chemistry and manufacturing R&D is done in-house. Due to its chemical businesses, LG Chem also has proprietary materials and processes. LG Chem is budgeting $1 billion in R&D for rechargeable batteries over five years.

Next Steps for A123

Spun out of the Massachusetts Institute of Technology in 2001, A123 makes lithium-ion batteries for products such as BAE Systems hybrid buses, Black & Decker power tools, Tesla electric sports cars, and utility grid-storage systems for utilities. It has plants in the United States, as well as in South Korea and China.

A123 secured over $100 million in incentives from Michigan and $250 million in federal Recovery Act funds to build a factory in Livonia, Michigan. The company also raised $400 million in a 2009 initial public offering. A123's Livonia plant began producing prismatic cells in June 2010 and has capacity to make batteries for 30,000 plug-in vehicles. A bigger campus is under construction in Romulus, Michigan that will produce everything from coatings to cells and packs.[53]

According to James M. Forcier, A123's vice-president of automotive solutions, the big challenge now is to sell enough batteries to make the U.S. production profitable. It will take up to five years for the industry to cut lithium-ion battery costs in half and before electric vehicles cost roughly the same as gas-powered cars, he said. Half of those savings will come from engineering improvements, but the other half must come from higher production volumes.

It is unclear that consumer demand will be sufficient to sustain the U.S. advanced battery industry. Mr. Forcier said. It takes up to $300 million to build one lithium-ion plant to supply batteries for 20,000 to 30,000 plug-in or electric vehicles. While government loans, rebates, and incentives will remain necessary for several more years, what U.S. manufacturers really need is help boosting demand. One "huge opportunity to help stimulate demand" is to electrify the big military and government vehicle fleets, Mr. Forcier said.

G. THE UNIVERSITY ROLE

Further technology advances and higher production volumes are needed to really push down costs and boost performance of advanced car batteries. In her presentation, Anna Marie Sastry of the Advanced Materials Systems Laboratory at the University of Michigan and CEO of the Ann Arbor-based advanced battery developer Sakti3, estimated that battery densities of around 500 watt hours per kilogram are needed in order to "see large degrees of electrification" of vehicles.[54] She predicted that when output of electric-car batteries hits 300,000 units a year, the price of lithium-ion fuel cells should drop from around $500 apiece now to $100 and meet the crucial threshold of around $300 per kilowatt.

Dr. Sastry noted that the University of Michigan is one of the first universities in the U.S. to invest in research and education aimed at improving lithium-ion cells and battery packs. It is collaborating with GM and the U.S. Advanced Battery Coalition to address all aspects of the electric power train. An

---

[53] See the summary of remarks by Jason Forcier of A123 in the next chapter.

[54] Kilowatt hours per kilogram are a measure of thermal heat capacity. Current lithium-ion batteries for vehicles tend to have a capacity of around 145 kilowatt hours. The current U.S. Advanced Battery Consortium target is to reach 300 kilowatt hours.

Energy Systems Engineering program founded by Dr. Sastry in 2007, meanwhile, has grown from nine students in 2007 to more than 200.

She reported that work also is accelerating at the Advanced Materials Systems Laboratory, which develops reliable algorithms for controlling and predicting battery performance under various conditions. The center coordinates more than 70 researchers from partners such as the DoE, National Science Foundation, LG Chem, GM Mainz Kastel, and Oak Ridge National Laboratories, and Ford.

As the market emerges and technologies develop, there is greater impetus for national laboratories, industry, universities, and government agencies to collaborate, Dr. Sastry said. "The technology pain is intense right now," she said. "A combination of mechanics, thermal effects, heat transfer, kinetics, and a whole host of other disciplines are required to build simulations that allow us to say how long a battery cell will live and how well it will cycle."

H. DEVELOPING A U.S. SUPPLY CHAIN

To be competitive in electrified vehicles, the United States also requires a domestic supply base of key materials and components. Ms. Gioia of Ford noted that electrified cars need special motors, transmissions, brakes, chargers, and devices that convert alternative current to direct current, for example. In

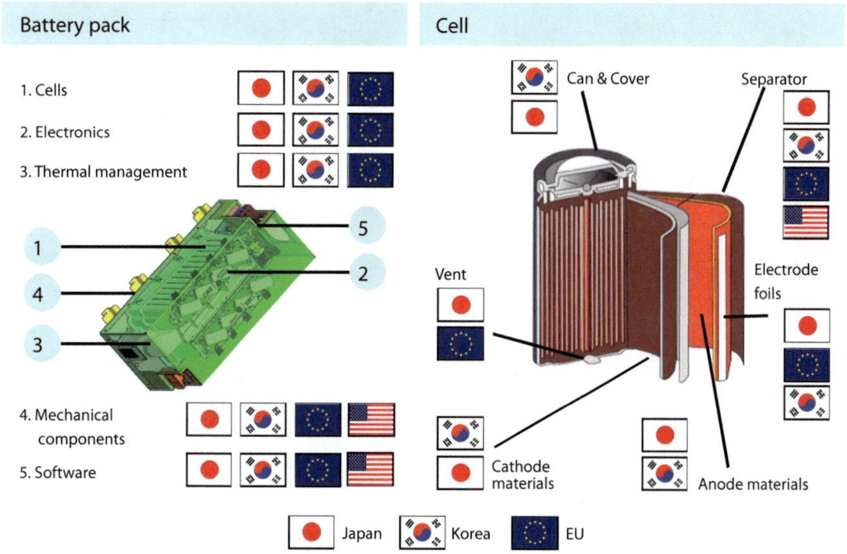

FIGURE 4 Most of the key supply base is in foreign countries.
SOURCE: Tom Watson, Presentation at July 26-27, 2010 National Academies Symposium on "Building the U.S. Battery Industry for Electric Drive Vehicles: Progress, Challenges, and Opportunities."

some cases, the components industry is underdeveloped. Chargers for electric vehicles are "ridiculously expensive today" and are made by "what was a cottage industry," she said. "We need main stream companies jumping into that."

Currently, U.S.-based battery plants must also import conductive materials, foils, separators, electrolytes, and other essential ingredients. Tom Watson, vice-president of technology at Johnson Controls Power, explained in his presentation that when his company launched a lithium-ion battery manufacturing joint-venture with France's Saft Advanced Power Solutions, it went "around the world to find what we believe are the best suppliers," At the venture's plant in France, which supplies hybrid battery systems to Daimler and BMW, cells, separators, and cathode materials "pretty much are coming out of Europe, Japan, and Korea," Much of the software and mechanical-component supply base for packs also is offshore, he said.

Johnson Controls-Saft is building a new lithium-ion batteries battery plant in Holland, Mich. The venture received a $299.2 million Recovery Act grant and $168.5 million in incentives from Michigan. When it looked for domestic sources of parts and materials, it found "a lack of a supply base here in the U.S.," Mr. Watson said.

Mr. Watson said that Johnson Controls wants to help develop domestic suppliers for North American plants, in part because it views creating local jobs and economic growth as part of its corporate responsibility. It has required each of its materials suppliers to build U.S. factories to process material. "We would really like to encourage a great mix of vertical integration in the U.S," he said. To this end, he noted that Johnson Controls is collaborating with start-ups, Argonne and Oak Ridge national laboratories, and universities to develop new materials.

The Business Case for Domestic Supplies

There also are compelling logistical reasons for sourcing lithium-ion battery supplies in North America. In his presentation, Michael E. Reed of Magna E-Car Systems observed that the complexity of the supply chain "adds significant cost" in the manufacture of advanced batteries.[55] Shipping materials from Asia is expensive and time-consuming. Regulations for handling equipment and hazardous materials vary from country to country. Companies must carry substantial inventories in case of supply glitches. Language barriers and time zones make communication difficult. Also, because many Japanese suppliers of crucial materials are controlled by large *keiretsu* business networks, U.S. battery makers often lack access to the latest technology, he said.

---

[55] Maga E-Car, based in Auburn Hills, Mich., is a unit of $17.6 billion Magna Steyr, a top tier-one supplier to the auto industry. Among other things, it buys lithium-ion cells from many companies and assembles them into a range of battery packs for customers

Establishing a domestic R&D and supply base for materials will make U.S. battery producers more competitive, Mr. Reed said. "But we have a lot of catch-up to do to become a viable competitor in this market," he said. The impressive investment in North American lithium-ion cell production since 2008 has not "been balanced by necessary investment in the supply chain itself," he said.

Several factors are holding up such investment, Mr. Reed said. Hybrid and electric-car production volumes are too small to justify investments in materials and parts plants. "Very few people are announcing programs in the tens of thousands of vehicles per year or higher." Each auto maker has its own standards and specifications. It can cost up to $3 million to fully develop a cell for a single customer and $10 million for a battery pack. For suppliers, the cost of developing and validating products for so many small programs "is really prohibitive," he said.

Ensuring a Secure Lithium Supply

Fears about a reliance on imported lithium and other key battery raw materials from China and elsewhere appear to be overblown, according to Linda Gaines of the Center for Transportation Research at Argonne National Laboratory[56]. In fact, the U.S. may be able to supply all of its own needs if reserves in California and Nevada are mined, technology improves at a reasonable pace, and battery recycling becomes common. Argonne developed models for a variety of scenarios, including a "maximum electric" scenario in which hybrid cars account for 25 percent of the U.S. market by 2025 and plug-ins account for 60 percent by 2050.[57] It also studied four battery chemistries requiring lithium and assumed the average battery would have a 100-mile range and weigh 500 kilograms.

Based on these assumptions, 50,000 to 60,000 tons of lithium will be used in electric cars on American roads by 2050—roughly twice current global production. But many new lithium mines are under development. In addition to major mines in South America, Dr. Gaines noted that industry analysts say some 100 U.S. companies are exploring for lithium at 150 U.S. sites that should meet domestic demand, she said.

Recycling of batteries and reuse of lithium, meanwhile, could sharply lower requirements for virgin lithium to less than 15,000 tons by 2050, well below current global production. Reducing the size and boosting the power output of lithium-ion batteries would slash needs even further.

---

[56] The U.S. industry relies on imported lithium, and some analysts predict demand from electrified cars will outstrip supply. See William Tahil, "The Trouble with Lithium: Implications of Future PHEV Production for Lithium Demand," Meridian International Research, December 2006 (http://tyler.blogware.com/lithium_shortage.pdf)

[57] Phil Peterson, Margaret Singh, Steve Plotkin, and Jim Moore, "Multipath Transportation Futures Study: Results from Phase 1," Office of Energy Efficiency and Renewable Energy, U.S. Department of Energy, March 9, 2007 (http://www1.eere.energy.gov/ba/pba/pdfs/multipath_ppt.pdf

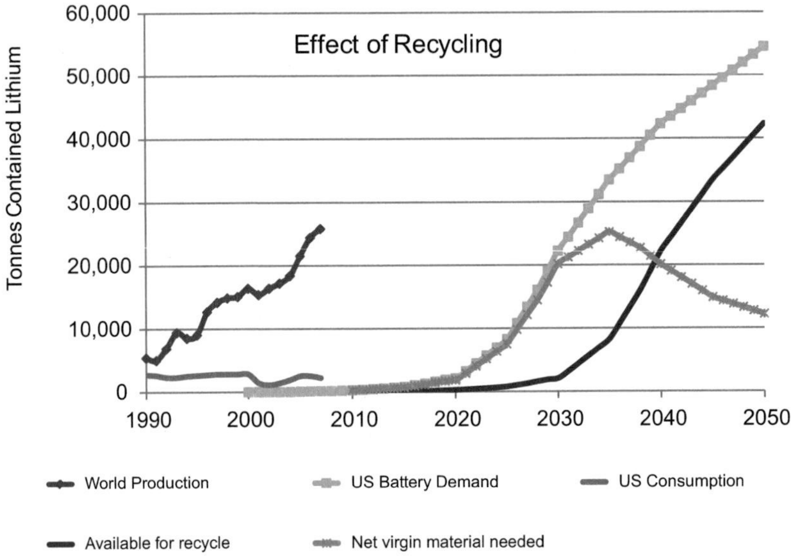

FIGURE 5 Recycling can drastically reduce virgin lithium demand
SOURCE: Linda Gaines Presentation at July 26-27, 2010 National Academies Symposium on "Building the U.S. Battery Industry for Electric Drive Vehicles: Progress, Challenges, and Opportunities."

Under aggressive projections of electric-car usage worldwide and conservative estimates of battery performance, lithium demand would reach 450,000 metric tons in 2050, Dr. Gaines noted. But if one uses smaller batteries in the projection and factors in recycling, projected demand drops to 100,000 metric tons, she said. The U.S. Geologic Survey estimates reserves in current mines around the world at 9.9 million metric tons and total world reserves at 25.5 million metric tons.[58] "It is not unreasonable to assume you can increase current world production by a factor of four in 40 years," she said.[59]

I. UNDERSTANDING THE CONSUMER AND MARKETS

The future of battery depends on the kind of cars that American consumers will want to drive. Many industry assumptions may be off base, contended Daniel Sperling of the University of California at Davis in his

---

[58] Data: U.S. Geological Survey, revised January 2010 data. See (http://minerals.usgs.gov/minerals/pubs/commodity/lithium/mcs-2010-lithi.pdf)

[59] Dr. Gaines noted that the Department of Energy has awarded a $28.4 million grant to Chemetall Foote Corp. to produce lithium at its operation in Silver Peak, Nevada. It also has awarded grants to several U.S. recycling companies.

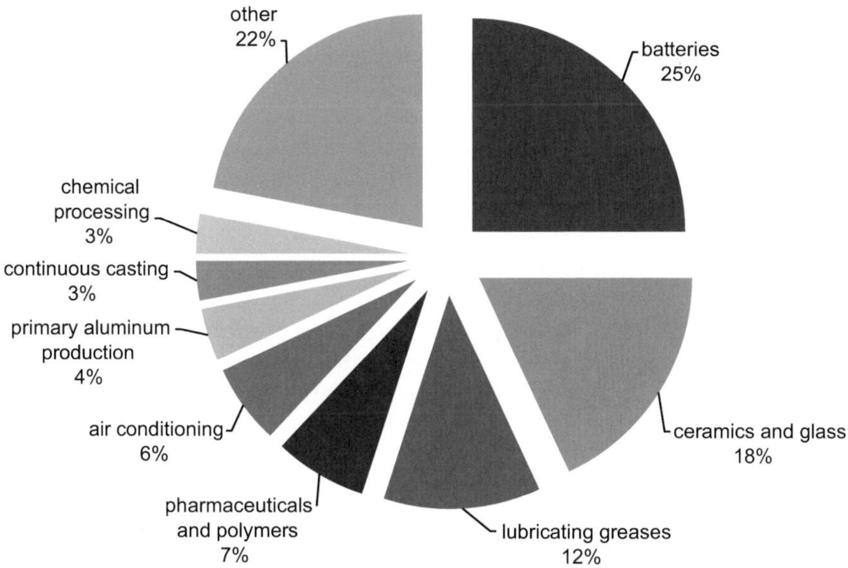

Source: SQM, cited in 2007 USGS Minerals Yearbook.

FIGURE 6 Batteries made up 25 percent of lithium use in 2007.
SOURCE: Linda Gaines, Presentation at July 26-27, 2010 National Academies Symposium on "Building the U.S. Battery Industry for Electric Drive Vehicles: Progress, Challenges, and Opportunities."

presentation. Dr. Sperling is founding director of the Institute of Transportation Studies, which has studied such questions for decades.

One important finding of the institute's research is that U.S. consumers appear to be satisfied with hybrid and plug-in cars with lower performance metrics than most "engineering experts" assume, Dr. Sperling said. In a study of BMW Mini E drivers, only one in six said the 100-mile driving range without recharging "was really a problem for them," he said. Most found ways to adapt. Another finding is that drivers rarely use public charging facilities, and few were interested in charging them at work. Instead, most drivers charge their cars at home at night.

Rather than being obsessed with performance standards and driving range, Dr. Sperling said research suggests consumers are more motivated by the "positive attributes" of electrified vehicles. These include energy independence for the U.S., helping the environment, avoiding gas stations, and not having to give their money to big oil companies and Middle East nations. "Electric vehicles give access to a whole new set of values and benefits for consumers," Dr. Sperling said. What's more, the more experience drivers have with electric vehicles, the more they like them. "People are remarkably willing to adapt to changing conditions and constraints if they see some value in doing so," he said.

Such findings, though tentative, suggest that "existing battery performance seems to be adequate for high market penetration by plug-in hybrid cars," Dr. Sperling said. "If we continue to follow the path we're on, trying to create an electric vehicle that is analogous to a gasoline vehicle, we are doomed to failure." Big government investments in public charging infrastructure also may be unnecessary.

GM's Focus on the Big Picture

Gary Smyth of General Motors agreed that understanding the actual consumer market and stressing environmental benefits are important if the electrification of transportation is to succeed. Instead of "niche plays," the industry must focus on transforming vehicle fleets. "You really have to look at what personal transportation in the future will be," said Dr. Smyth, executive director of GM's North American R&D laboratories.

Stressing economic benefits of hybrid cars alone won't suffice because the actual fuel cost savings "are really quite limited" considering the $3,000 to $6,000 added cost of buying one, Dr. Smyth said. If a mid-sized hybrid saves 30 percent to 40 percent on gas and is driven 12,000 miles a year, a family saves just $300 annually on fuel. Even if gas is $6 a gallon and one looks at so-called "third generation" electric cars expected by 2025, the savings aren't huge, he said.

One conclusion from research is that consumer needs will depend very much on where they live, he said. "What you need for the mega cities and hyper cities is very different from what you need in Texas and the Midwest," he said. "It really is about a portfolio of solutions." For many U.S. families, "range anxiety" is a real issue, a lesson Dr. Smyth said GM learned from its experience with the EV1[60] in the 1990s.

GM, therefore, is determined not to "compromise on the utility of the vehicle for the customers," Dr. Smyth said. It sees the Volt, a cross between a plug-in hybrid and a pure electric vehicle that can run 90 percent of the time on electricity, as a learning exercise, he said. Although the 400-pound, six-foot-long battery pack has passed road tests with flying colors, he suggested it is too big and heavy.

The one given is that future transportation solutions will have to be low-carbon, he said. With federal financial help, GM is building a plant for Volt batteries in Michigan's Brownstown Township. GM also invested $246 million in motor and electric-drive facilities and received $105 million in federal funds for a plant in White Marsh, Maryland, to produce high volumes of electric

---

[60] The EV1 was produced by General Motors from 1995 through 1999. After being rolled out in several U.S. cities, GM cancelled the program because it determined the vehicle would not be profitable.

motors starting in 2013. In addition to lithium-ion technologies, GM also must work on fuel cells and hydrogen power, which will be commercially viable by around 2016, Dr. Smyth said.

Electrifying the Trucking Industry

While the future of hybrid and plug-in cars is under debate, the electrification of America's trucking fleet is making clear progress. Bill Van Amburg of CALSTART, an industry organization promoting clean transportation, said in his presentation that hybrids now account for about 40 percent of the new market for transit busses. Companies from FedEx to Coca-Cola are also introducing hybrid trucks to their delivery fleets. Major builders of regional and long-haul trucks such as Navistar, Freightliner, Kenworth, and Peterbilt have electrification programs.

The unit volumes are small—there are only 2,000 hybrid trucks on U.S. roads spread over many market niches. But sales are doubling every year, Mr. Van Amburg said. "We are seeing real movement, a real transition, in the truck world to advanced technologies." According to estimates calculated by CALSTART, 30 percent of the U.S. truck market will be ripe for hybrid technologies by 2020. Because trucks require many more battery cells than cars, moreover, this market segment will be important to lithium-ion battery makers.

A major reason for the conversion is that the business case for trucks is more clear-cut. Because trucks burn so much petroleum-based fuel, investments in hybrids can sometimes pay themselves off in three to five years, Mr. Amburg said. Fleet owners are starting to demand electrified vehicles. Federal and state policies also provide incentives to the truck industry to adopt electrification. For example, tougher air pollution standards have prompted Los Angeles to develop a zero-emission highway corridor for trucks carrying freight from its seaports.

The experience of the trucking industry illustrates "the power of public-private partnerships" in moving R&D into early production, Mr. Van Amburg said. One quarter of CALSTART's 130 corporate members are in the Midwestern "manufacturing corridor, he said. CALSTART is part of the Hybrid Truck Users Forum, in which commercial trucking and manufacturing companies discuss ways to deploy advanced technologies for specific applications. Mr. Van Amburg estimated such forums and working groups sped up adoption of electrified trucks by two to five years. CALSTART collaborates with the U.S. Army's TARDEC on strategies to deploy green technologies in military vehicles.

CALSTART also manages a $20 million tax-incentive scheme in California. California pays half of the incremental cost of buying a hybrid instead of a conventional truck. Some 600 trucks were purchased through the program, increasing hybrids on the road by 30 percent, he said. The trucking industry offers is a good case study of "how we might get things moving" in the U.S. in transportation electrification, Mr. Van Amburg said.

J. UPGRADING THE WORKFORCE

The need for a trained workforce is essential for the United States to compete globally in advanced-technology industries. "I think everyone here knows the country is struggling with a K-12 education system that is weak," commented Bill Harris of Science Foundation Arizona.[61] Moreover, there are concerns the U.S. is not training enough engineers to support a large advanced battery industry. Dr. Sastry of the University of Michigan cited warnings by the Institute of Electrical and Electronics Engineers' Power & Energy Society that electrical power engineer graduation rates don't meet the nation's current and future needs.[62] "We're lacking the people to do this," Dr. Sastry said. Scientists and engineers are "an absolute requirement for a sustainable business," she said.

Even if there is an influx of new engineering students, universities and colleges may not be ready to train them. The IEEE estimates that within five years, 40 percent of full-time U.S. senior engineering faculty will be eligible for retirement.[63] The lithium-ion industry also needs new kinds of engineers fluent in physics, electro-chemistry, and even biology, several speakers noted. As the U.S. Army's Dr. Pellegrino noted, engineering students need experience early on in working in multidisciplinary teams.

Michigan's Workforce Training Push

To address the need for a qualified labor force, Michigan has perhaps "the most aggressive workforce training program in any state," said Andy Levin, acting director of the state's Department of Energy Labor, and Economic Growth. Michigan's No Worker Left Behind Initiative, launched in 2007, offers $5,000 a year or $10,000 for two years of college or university tuition to any person who is unemployed, about to be laid off, or has a family income of less than $40,000. More than 135,000 Michigan workers have already gone through the program to earn or finish associate, bachelor, or master's degrees.

Mr. Levin explained that rather than training new workers, the goal of the program is to upgrade the skills of existing workers. The state began focusing on labor needs for electric vehicles because it saw green-technology industries as a potentially major new employer. Between 2005 and 2008, overall employment in Michigan's private sector shrank by 5.4 percent, he noted. In

---

[61] See National Research Council, *Rising Above the Gathering Storm: Energizing and Employing America for a Brighter Economic Future*, Washington, D. C.: National Academies Press, 2007.

[62] Amy Fischbach, "Engineering Shortage Puts Green Economy and Smart Grid at Risk," Transmission and Distribution World, April 21, 2009. (http://blog.tdworld.com/briefingroom/2009/04/21/engineer-shortage-puts-green-economy-and-smart-grid-at-risk).

[63] U.S. Power and Engineering Workforce Collective, "Preparing the U.S. Foundation for Future Electric Energy Systems: A Strong Power and Energy Engineering Workforce," IEEE Power and Energy Society, April 2009.

contrast, green employment grew by 7.8 percent, adding 2,200 new jobs, 700 of them in companies that did not exist in 2005.

Drawing on advice from GM, Ford, Chrysler, Japanese automakers, as well as from university research, Michigan officials realized the electric-vehicle sector needs labor with different skills than the traditional auto industry. In response, as part of a $6 million green jobs initiative, state agencies formed "skills alliances" with employers. The Michigan Emerging Market Skills Alliance, for example, works with small tool-and-die suppliers that must diversify. The Michigan Academy for Green Mobility, meanwhile, trains engineers for vehicle electrification.

Mr. Levin noted that Wayne State University and Michigan Technological University lead the electric vehicle programs, which so far have trained 300 workers. The state is also talking to battery manufacturers about a similar program.

Wayne State's New Degree Programs

Furthermore, Wayne State University is developing a comprehensive degree program for electric-drive technology and batteries with Department of Energy funding. The program's advisory board includes Ford, TARDEC, and Compact Power.

Wayne State has dubbed its program E3, standing for "electrification, the economy, and education," explained Simon Ng, director of the school's alternative energy technology program. The school offers a master's degree in electric-drive vehicle engineering, a bachelor's in electric transportation, and associate degrees in automotive technology and electronic engineering technology. It also offers an undergraduate concentration and a graduate certificate in electric-vehicle engineering.

To design the curriculum, Wayne State received input from auto makers and parts suppliers, Mr. Ng explained. It also studied best practices in electric vehicle-related curricula from around the world. Mr. Ng recently visited key Chinese universities. "I am really glad we have a complementary program now in the United States to do similar things," he said. "Otherwise, we would be falling behind."

The program aims to be comprehensive, industry-oriented, and to make a national impact, Dr. Ng explained. The curricula, therefore, includes electrical, mechanical, chemical, industrial engineering and alternative-energy technology courses. The program also stresses real-life laboratory experience at the university and at companies. Wayne State wants to extend the program's reach through distance-learning. Out-of-state students would conduct laboratory experience through simulations, remote controls, and week-long visits to the Detroit campus, he said. Wayne State is adding laboratories for fabricating new materials and cells, electric controls, characterization of battery packs, and electric-drive propulsion systems.

Indiana's "Middle Skill" Focus

Battery companies also are helping develop the skilled workforce. Indianapolis-based EnerDel, the lithium-ion solutions unit of Ener1, is working with Ivy Tech, an Indiana community college that has 23 campuses and 130,000 students.[64]

One of Ivy Tech's strengths is working with industry to train what EnerDel Chief Financial Officer Robert Kamischke described as "middle skill workers," those with two years of college but short of a bachelor's degree in engineering. Such workers are in short supply. Fifty-six percent of demand for all workers in Indiana is classified as middle skill, he noted. Only 45 percent of the state's workforce has sufficient training.[65]

Five out of six jobs in the advanced battery industry will require middle- to high-skill workers, Mr. Kamischke said. EnerDel develops and manufactures lithium-ion battery solutions for consumer electronics, transportation, and power-generation companies and the military. Customers include Nissan, Volvo, TARDEC, and Think Automotive. It received $118.5 million through the Recovery Act to build a plant.

For demanding cell and electrode fabrication processes, EnerDel will seek workers with two-year applied sciences agrees, he said. Such skilled workers also are needed in emerging-technology industries such as wind turbines, solar panels, and renewable-energy power plants.

In all, Mr. Kamischke predicted that Ivy Tech "will be part of the backbone of building this emerging middle work force for the renewables age." The school offers an associate's degree in applied science with focuses on industrial technology, advanced manufacturing, and engineering technology. For the transportation sector, Ivy Tech is developing curricula for the electric-vehicle, recycling, and first-responder industries with a Department of Energy grant. It also is developing a certificate program for electric transportation technicians.

[64] "EnerDel's parent Ener1 declared bankruptcy in January, about seven months after Norwegian EV maker Think, in which Ener1 was an investor, did the same. EnerDel restructured in March [2012] after Ener1 received $86 million in new equity and debt-holder agreements." Most recently, EnerDel gave Purdue University's College of Technology "a collection of lithium-ion battery cells and research data worth about $263,000. The gift complements the $4.7 million grant that the university and Ivy Tech Community College got from the U.S. Department of Energy to advance training geared towards the electric-energy industry." See Danny King, Autoblog Green, October 1, 2012. Access at http://green.autoblog.com/2012/10/01/enerdel-battery-business-purdue-li-ion/.

[65] Data from Indiana Department of Workforce Development and U.S. Census Bureau.

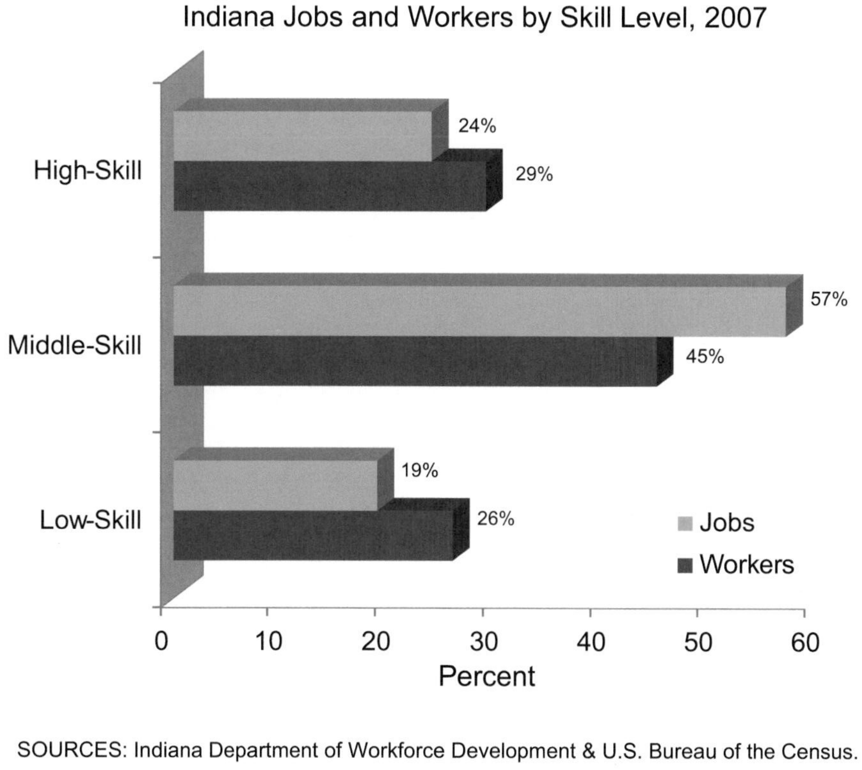

FIGURE 7 Supply and Demand for Middle-Skill Jobs in Indiana
SOURCE: Robert Kamischke Presentation at July 26-27, 2010 National Academies Symposium on "Building the U.S. Battery Industry for Electric Drive Vehicles: Progress, Challenges, and Opportunities."

K. INITIATIVES IN RESEARCH AND DEVELOPMENT

Growing Activities at the DoE

Speaking at the conference, David Howell, who leads the hybrid electric systems program at the DoE's Office of Vehicle Technologies said that the Department of Energy expects to continue ramping up its efforts in advanced batteries. The Obama Administration requested a sharp funding boost for energy-storage research, to $209.7 million in FY 2011, with a greater focus on grid storage. The transportation research budget would rise by $20 million. Other federal battery-related R&D programs also have grown.

Plug-in hybrids currently are the main focus of the DoE-led U.S. Advanced Battery Consortium, explained David Howell. One goal is to push pulse power discharged by plug-in hybrid batteries from around 25 kilowatts in 2010 to 38 to 50 kilowatts by 2015 and to 80 kilowatts for all-battery electric vehicles by 2020. Prices for plug-in hybrid batteries "seem to be on track" to drop from around $2,500 now to $1,700 in 2012. "But when you go to higher mileage plug-ins and electric vehicles, the targets get a lot tougher," he said. "So we have to move on to the next generation of lithium ion chemistries or beyond to meet the targets." He noted that A123, Johnson Controls-Saft, EnerDel, 3M, and other companies are completing DoE-funded battery R&D projects, and that 12 new projects are being negotiated.

Dr. Howell also noted that improved materials also are receiving greater attention. The Vehicle Technology Program and Oak Ridge National Laboratories support five companies working on materials and processing technologies. In 2009, the DEPARTMENT OF ENERGY awarded several grants to companies working on advanced anode materials. Future research projects, he said, will focus on high-capacity cathode materials, high-voltage electrolytes, and lithium materials. The Vehicles Technologies Program also funds extensive research into many areas of electro-chemical cells, with $34 million a year going into 60 projects at 10 national laboratories and 12 universities

Finally, Dr. Howell noted that DoE's ARPA-E program awarded 10 new grants for breakthrough research in 2010. These awards include projects in lithium-air batteries at the Missouri University of Science & Technology, an all-electron battery at Stanford, and high-performance and ultra-low cost rechargeable batteries at MIT. "Even if half of these research projects are successful, "that would be a big win for us," Mr. Howell said.

Budgets also have risen for DoE's Basic Energy Sciences project, which focuses on fundamental materials and electrochemical process research. Five of the 46 Energy Frontier Research Centers funded by the project do work related to batteries and vehicle technology, Mr. Howell noted. The grid-storage budget of the DoE's Office of Electricity, meanwhile, rose from $3.6 million in FY 2009 to a requested $40 million in FY 2011.

Military Battery Research Programs

The Defense Department also is boosting battery research. The U.S. military has invested $150 million over the past six years in R&D in areas like Silent Watch, Silent Mobility, power for soldier communications, and pulse power for armor—all of which require advanced storage, explained Sonya Zanardelli, TARDEC's energy-storage team leader. It also invests in alternative chemistry, new material, and thermal management research.

TARDEC alone has 60 research projects underway in energy storage, Ms. Zanardelli said. They encompass basic research, applications, manufacturing processes, and battery management and safety. The Army wants to replace nickel-zinc batteries with lithium-ion for starting, lighting, and ignition systems for combat vehicles, for example. TARDEC also is exploring large-format lithium-ion phosphate and nickel cobalt oxide batteries that are lighter, run longer, and offer greater temperature range.

TARDEC has a number of manufacturing technology programs aimed at cutting cost and enabling high-volume manufacturing, Ms. Zanardelli explained. A project started in 2004 for future combat systems focused on automating lithium-ion production processes and halving their cost. TARDEC is applying the knowhow in light-tactical vehicles, she said.

TARDEC collaborates with units across the DOD and other federal agencies to share knowhow and minimize overlap, Ms. Zanardelli said. The Army also is developing lighter-weight batteries for soldiers, for example, and the Air Force is developing hybrid systems for unmanned aerial vehicles that operate 40 to 50 hours and need thousands of watts of power. The U.S. Navy is looking to use hybrids for unmanned underwater vehicles, shallow-water combat submersibles, submarine small distributed power systems, and surface ship fuel economy.

Kentucky's Advanced Battery Manufacturing Center

The state of Kentucky is another state that is seeking to play a role in advanced battery manufacturing by establishing a new R&D center with Argonne National Laboratory. The Kentucky-Argonne Battery Manufacturing Research and Development Center, based at the University of Kentucky, is preparing to erect a new laboratory building with $10 million in funding from NIST and $4 million from the state.

Ralph C. Brodd, the director of the Kentucky-Argonne center, noted in the conference that a key mission is to develop new manufacturing processes and lines for advanced batteries. The overall mission, he noted, is to re-establish the United States as a world leader "in manufacturing technology and capability." To cut U.S. dependence on imported cells and equipment, he said, "there need to be new concepts and processes to produce batteries more efficiently at lower cost."

The Kentucky-Argonne center aims to accelerate production of advanced technologies from national laboratories and universities, Mr. Brodd said. It expects to design new cell fabrication processes for both the cylindrical and prismatic formats that boost speed, density, and cycle life. It also will facilitate national interactions among industry, universities, and National Laboratories to "optimize a good supply chain and develop a viable battery manufacturing industry here in the U.S." Mr. Brodd said.

Another goal of the center is to develop a roadmap identifying the infrastructure and technology elements "required to develop and maintain a leadership position that we feel we absolutely must generate," Mr. Brodd said. These efforts, he added, also could boost Kentucky as cost-competitive manufacturer of cells for the global market.

The Role of the Manufacturing Extension Partnership

American battery manufacturers also can tap the extensive resources of the Manufacturing Extension Partnership, said David C. Stieren, who oversees technology deployment at the MEP. This "federal-state-private partnership" that aids manufacturers is managed by the National Institute of Standards and Technology, which funds programs working on research, performance characterization, and measurement methods for battery technologies. NIST also administers the Technology Innovation Program, which awards grants in the battery sector, he noted.

The MEP works with "companies that want to be proactive, want to expand, and want to establish their niche in the marketplace," Mr. Stieren said. Services are delivered through MEP's network of 60 centers, which are found in each state and have 1,600 staff that interacts daily with manufacturers. The MEP also contracts with 2,300 service providers. Staff can tap their nationwide network of contacts in industry, National Laboratories, government agencies, and universities to help manufacturers find technology, funding, suppliers, training programs, or potential customers, he explained. The MEP works some 31,000 companies each year. "We really have a fantastic reach to the nation's manufacturing base," he said.

In the battery industry, the MEP engaged in 120 projects with companies across the U.S. between 2005 and 2009, Mr. Stieren said. The projects involved 47 different companies in 26 states. Roughly one-third had 50 employees or fewer. About half had more than 100 employees.

The MEP helps battery manufacturers with myriad challenges. They include Six Sigma quality, marketing, road-mapping, lean manufacturing, energy efficiency, export market access, supply-chain management, and product development, Mr. Stieren said. These battery projects are credited with helping generate $69 million in sales, $35 million in cost savings, $32 million in investment, and 1,041 new or retained jobs.

L. THE ROAD AHEAD

Michigan's Next Steps

Now that Michigan has enticed battery manufacturers to set up factories in the state, the MEDC is reassessing its "economic tool kit" to promote the next phase of development, Eric Shreffler, who leads the MEDC's advanced energy storage program, said in his presentation. It also will have to work with a new set of policymakers: Two-thirds of Michigan's legislature will turn over in the fall 2010 elections, and a new governor will be elected. The MEDC also plans to spend more time in Washington urging lawmakers to keep moving the advanced-battery industry forward.

The MEDC is focusing on building out the advanced-battery supply chain in Michigan and exposing companies to complementary markets, Mr. Shreffler said. It also will work to strengthen what the MEDC calls "the alliance." The state will attempt to better align its initiatives in batteries and advanced materials with the priorities of federal agencies and national laboratories, he said, to improve the scope "for Michigan companies to plug into federal opportunities," he said. By collaborating on research and commercialization of "dual-use" technologies, the state can create more opportunities to generate and retain jobs, he said.

One example of such state and federal collaboration is a new $27 million, three-year joint program involving Michigan, Oak Ridge National Laboratories, and TARDEC to commercialize advanced-storage and light-weight material research in Department of Energy laboratories and adapt it for military use, Mr. Shreffler said. By demonstrating that such approaches work, the MEDC hopes to raise further funding for such "dual-use" projects.

The main challenge now is execution, Mr. Shreffler said. "We have to execute as an economic development agency," he said. "Our cell manufacturers and suppliers must execute to build out their capacity. And the federal government has to execute by not abandoning the path that we've gone down."

The Growing Market for Electrified Vehicles

Where America's nascent battery industry goes next was the key concern raised by industry and economic-development officials in the symposium. The industry has gone through the initial learning stage of R&D, Dr. Smyth of GM said. Now comes the commercialization stage. "It is the Valley of Death," he said. "And it won't be a narrow valley." To sell electric vehicles, car makers must make them affordable, offer the right technologies to consumers, and develop the supply chain, he said. While the U.S. now is installing manufacturing capacity, "the knowledge to build the equipment, set the details, and design the processes for the future is not being brought here yet," said Ms. Gioia of Ford. Without that, the U.S. battery industry will still trail Japan, South Korea, and China.

Economic development officials also said they recognize that enticing companies to set up factories with subsidies was the easy part. The question is where to go next. "That is something that we as a state are really very concerned about," said Gary Krause, the MEDC's director of federal partnerships and initiatives. Michigan has "literally bet the farm" on the electrified vehicle industry as a means of diversifying its economy, he said. "There is $6 billion in state, federal, and private investment on the table. That is a lot. So the issue of completing this task from a policy standpoint really is key."

One message that emerged is that policymakers will have to be patient. Even under the most optimistic scenarios, the vast majority of new cars sold in America for several decades will be gas-powered, several speakers pointed out. It then will take years before owners must replace those vehicles. "If you are trying to realize maximum benefit out of a new technology that is introduced today, it takes three or four decades to get to that point," Mr. Davis of the Department of Energy said.

To survive the next four or five years, however, U.S. advanced battery manufacturers will have to be able to sell large volumes of batteries. America's ability to export significant volumes, several speakers said. Dr. Charles Wessner of the National Academies noted that many other nations have industrial policies that favor domestic production and discourage imports. "Most countries are willing to export to us, but the other way is harder," he said. Mr. Forcier of A123 said policies in promising markets like China and Germany strongly favor domestic production. "European business will be won and made in Europe, and Asian business will be won and made in Asia," he said.

M. THE ROLE FOR POLICY

As we see below, several speakers at the conference suggested how federal policy could advance a U.S. advanced battery industry beyond support for research and manufacturing.

Early Procurement to Boost Demand

Les Alexander, A123's general manager for government solutions, noted that federal priorities need to shift to "demand-driven stimulation rather than stimulating manufacturing and research. We can create the best battery in the world, but without vehicles to put them in this industry will go back overseas and we will have stimulated another country's industries."

One proposal is for the government to boost demand through purchases of electric vehicles for federal fleets. Senator Stabenow noted that the government will, as a symbolic measure, buy the first 100 Chevy Volts. "I would like to add a few zeros to that" and do the same for Ford and Chrysler, she said. Senator Stabenow also noted the federal government owns some 700,000 vehicles, including those operated by the U.S. Postal Service and the military. She noted that she is supporting a Senate bill that encourages federal agencies to buy electric vehicles.

In this regard, Mr. Amburg noted that the Advanced Vehicle and Power Initiative, a program backed by TARDEC, calls for replacing 8 percent of the government truck fleet annually with electrified vehicles.[66] The initiative, he said, could be "greatly beneficial to the truck world and be really helpful to light-duty manufacturing."

---

[66] The Advanced Vehicle and Power Initiative is an effort facilitated by TARDEC to advance collaboration among manufacturers, academia, and government to accelerate deployment of advanced vehicle technologies. A May 25, 2010, draft of AVPI's policy white paper is available on the CALSTART Website (www.calstart.org/Libraries/HTUF_Documents/AVPI.sflb.ashx).

Improving Government Incentives

As several executives and policymakers also observed at the conference, modifications in government incentives could also boost demand. In this regard, Mr. Reed suggested extending the length of time incentives are available given that "battery makers operate on a five- to seven-year time horizon."

Senator Stabenow recommended that the U.S. adopt a more formal system for longer-term financing for companies commercializing their technology, as do other nations. She noted that bills in the House and Senate call for establishing a Clean Energy Development Administration[67] that would help fund early-stage commercialization of new technologies.

Allowing buyers of hybrids and plug-ins to get $7,500 federal rebates at the time of purchase rather than as a tax refund could also stimulate more demand, Senator Stabenow said. The Cash for Clunker's program, which was "successful beyond my wildest dreams," used such an approach, she said. "That is more helpful than waiting until you fill out your taxes the next year."

Establishing Common Standards

Standards are another major question facing the advanced battery industry. Each automaker "has its own special set of requirements that drives the whole process," Mr. Reed of Magna E-Cars said. "Often, you have cell or pack technology that has been developed and qualified to one set of standards. But you may still need to spend millions of dollars to re-qualify it for another OEM[68]."

It may too early for the U.S. to set industry-wide standards for cell size and capacity, as has Germany. Based on his experience in the battery industry, Mr. Reed said, "this is something that is not going to happen by committee," he said. First, electrified vehicles must be produced in much higher volumes than they are now. That will determine the "winners in the survival-of-the-fittest process," he said.

Mr. Reed suggested that the U.S. government and industry could start instead by standardizing the way materials and cells are assessed. That way, potential suppliers "have a clear understanding of what the expectation is." If customers have consistent expectations, the costs of qualification and development "can be kept to a reasonable level," he said. Mr. Watson of Johnson Controls-Saft said he also does not see a rush toward standardized cells.

---

[67] Provisions for a Clean Energy Development Administration (CEDA), popularly referred to as a "green bank," to fund commercial-scale deployment of clean-energy technologies was included in Sections 184-190 of the American Clean Energy and Security Act of 2009 (H. R. 24540, which passed the House of Representatives on June 26, 2009. A similar institution, called the Clean Energy Deployment Administration, is included in The American Clean Energy Leadership Act (S. 1492) before the Senate.

[68] According to the Dictionary of IBM computing terminology, an Original Equipment Manufacturer, or OEM, is "a manufacturer of equipment that may be marketed by another manufacturer."

If there were common rules for charging, handling, and transportation standards, however, "the better off we all will be."

The Charging Infrastructure Question

The question of whether a national network of public-charging stations is required to foster wider consumer acceptance of electric cars is a major issue for federal and state governments interested in advancing vehicle electrification. As Senator Stabenow noted, it will not be enough to have electric vehicles on the road; "We have to make sure that the infrastructure in there as well."

Most speakers agreed some public charging facilities are needed to ease the so-called "range anxiety" of drivers who fear they will be stranded should their car batteries run out of power. Dr. Sperling of the University of California at Davis pointed out that few Japanese bought electric cars until a utility set up public charging stations—even though few drivers actually use them. "Public charging stations have psychological value," he said. The problem is that there "is no business model there because they won't be used very much." A minimal number of stations are needed at least in the beginning to address consumer anxiety, he said. "But it is not a key aspect of building up an electric vehicle industry."

There is "a fair alignment" among auto makers that public charging is a low priority, said Ms. Gioia of Ford. Dr. Smyth of GM agreed. Charging systems for homes are more urgently needed, they agreed, with charging stations at work sites and vehicle depots occupying the next priorities. Here, cost is a major issue. Home chargers for small, basic plug-in hybrids can be installed for less than $200, Ms. Gioia said. But all-battery electric charging systems cost around $2,000. Workplace or public stations can cost $50,000 each.

GM does agree that some public-charging infrastructure is needed "to make this comfortable for customers," Dr. Smyth said. GM is working with around 300 North American utilities to set up charging facilities.

Power Grid Concerns

America's electrical power grid is another infrastructure concern. Dr. Good questioned whether there will be enough generation capacity around the country to charge all vehicles, especially under the most optimistic scenarios of electric vehicle sales. She said she is "not sure adequate models have been developed" to account for an electric vehicle market penetration rate of 25 to 30 percent in a decade, rather than 5 percent as most analyst project now. Many parts of the U.S. currently do not have much excess capacity, she noted.

Responding to Dr. Good's question, Mr. Van Amburg cited analyses by U.S. power utilities that indicate there will be sufficient power in the grid because most cars will be charged at night, during off-peak hours. The bigger issue is making sure there is enough power in specific areas with high concentrations of electric vehicles, he said. Dr. Sperling agreed. First, it will be a "very long time" before 25 percent of cars will be electric. A more immediate concern is whether transformers must be upgraded in areas with high concentrations of electric vehicles, he said.[69]

The government's Smart Grid initiative should aid the rollout of electric vehicles, Mr. Davis of the Department of Energy said. The Office of Electricity manages a program that has invested more than $8 billion, both in federal and non-federal funds, in more than 100 projects. They include 100 plug-in hybrid charging stations, 176,000 load control devices, 206,000 "smart transformers" that allow for preventive maintenance, and 671 automated substations that account for 5 percent of the 12,466 transmission and distribution substations in the U.S., he said. Smart grid isn't essential for rolling out of electric cars in 2010, he said. "But when you start talking about a million vehicles, smart grid becomes very important pretty quick."

Dr. Good said current statistical models still don't seem adequate to allay concerns that the grid won't be able to support dramatic growth in vehicle electrification. "If you are trying to rev this up to 25 percent in the next 10 years, you had better get on that problem now," she said.

N. WHAT WE HAVE LEARNED

In the concluding roundtable of the conference, Mary Good asked the participants to offer thoughts on some of the lessons from the Michigan battery initiative.

Leadership from the State: Bill Harris noted that he was particularly impressed with the Michigan government's readiness to invest in the future and diversify the state's economy.

Capturing Regional Synergies: Mr. Harris noted that "Kentucky's goals and ambitions with Argonne match nicely with what is going on in Michigan, and there could be reasons to look at doing things together."

Learning Across State Lines: Mr. Harris further noted that "you need some legislators to understand what other states are doing. The absence of informed representatives hurts the dialogue."

Federal-State Partnerships: Mr. Les Alexander of A123 said that the coordination of state, federal, and military efforts remains important to drive development and deployment of the advanced battery industry.

---

[69] "Today, almost every major investor owned utility (IOU) in the U.S. is modernizing, or planning to modernize, its existing power distribution or transmission system or both. This is happening to prepare for expected changes, such as the adoption of renewable power and electric vehicles." See Farah Saeed, "What Does Grid Modernization Mean for the Economy and Job Growth?" Frost & Sullivan Principal Consultant. Access at the Electric Light and Power website at http://www.elp.com/index/display/article-display/1932717179/articles/utility-automation-engineering-td/volume-17/issue-3/departments/notes/miso-delivers-billions-in-benefits-to-region.html.

Importance of Demand: Mr. Alexander said that at this stage, demand-driven stimulation is more important than stimulating manufacturing and research. He warned that if electric vehicles are not built and purchased, "there is a risk that this industry will go away." He suggested that the electrification of military applications, postal fleets, and other government vehicles can help create this demand. Gary Krause of MEDC added that there also needs to be a cultural shift towards the acceptability of electric vehicles, including cars, large trucks, and other vehicles. He suggested a broad based educational effort that does not bear a heavy government fingerprint.

Incentives: Dr. Sastry of the University of Michigan stressed the importance of engaging "the next generation of companies and people." She suggested engaging student teams, education programs, and programs like the X Prize to spur innovation.

O. A MATTER OF COMMITMENT

An underlying concern voiced by many industry and policy leaders at the conference was that the political commitment needed to take the advanced-battery initiative to the next level may not be sustained over the longer term. As Senator Stabenow put it, many investors are still "sitting on the sidelines." She noted that for industry to make the large, long-term investments needed for the U.S. to be competitive, the direction of federal energy policy must be clear.

Speaking at the conference, Mr. Van Amburg observed that efforts such as that to electrify the U.S. trucking fleet will require "a coordinated set of standards, policy incentives, and regulations across the whole continuum to the market." While the U.S. does a good job at R&D, it has been "dropping the ball" when it comes to developing high market volumes "to justify the investment by the manufacturers and suppliers," he said.

Battery industry executives who spoke at the conference concurred that continued government financial help is essential as the industry further matures. Mr. Forcier of A123 said loans and incentives will probably be required for four to five years, until the costs of hybrid and plug-in cars approach those of gas-powered cars. Firm commitment by America's leading corporations also is essential, said Mr. Alamgir of Compact Power. Had U.S. companies and the government extended more financial help in the 1990s, as did those in Japan, more U.S. lithium-ion makers may have survived. What's needed are "gutsy and visionary leaders" in the U.S. private sector who "believe in the future of this industry and are committed to providing funds," he said.

America now faces a "great, once-in-a-lifetime opportunity" to emerge as a leader in advanced vehicle technologies, said Jim Greenberger, executive director of the National Alliance for Advanced Technology Batteries. "But it also is a tremendous responsibility. It is a responsibility of every one in this room to build an industry that is truly sustainable, to create jobs that are sustainable, and to make some real progress on moving our country away from petroleum dependence."

This conference report captures the views of state and federal officials as well leaders in industry and academia on the future of the advanced battery industry in Michigan. The next chapter provides detailed summaries of their remarks.

II

PROCEEDINGS

Day One

Welcome

Greg Main
Michigan Economic Development Corporation

Michigan is proud to host the symposium on developing the advanced-battery industry for electric vehicles and appreciates the partnership with the National Academy of Sciences, said Mr. Main, the Michigan Economic Develop Corp.'s president and CEO. Mr. Main recalled a comment President Barack Obama made at a recent groundbreaking ceremony for a new battery plant in Holland, Mich. "This event really signals where Michigan is going and where America is going," the President said.[1]

One reason Michigan has emerged as the U.S. center for advanced batteries is that it targeted the industry before the federal government and other states did so, Mr. Main explained. It also crafted a strategy to develop and grow the industry, he said. Michigan has long been the home of the auto industry. "Certainly that industry has been in difficult straits in the past few years--and the past decade, really," he said. "We saw this as an opportunity to help the auto industry diversify and to help Michigan diversify going forward."

In August 2009, the U.S. Department of Energy awarded $1.3 billion in Recovery Act grants to 13 Michigan-based projects to support advanced-battery and electric-vehicle manufacturing. That represented more than half of the federal funds devoted to the advanced battery initiative. "It was a great day in Michigan, but it came after more than three years of effort on our part, and certainly a significant effort on the part of the Big Three to move forward in this exciting new area," he said. Just since August 2009, "16 advanced battery and battery technology companies have committed to plants in Michigan. Those projects will create an estimated 62,000 jobs by 2020," he said. In addition to the LG Chem-Compact Power facility, "five other lithium-ion cell and battery plants were under construction [as of July 2010] or soon will be. They are by

---

[1] For text of President Barack Obama's speech at the groundbreaking of the battery plant by LG Chem and Compact Power, see "Remarks by President in Holland, Michigan on Investing in Clean Energy," the White House Office of the Secretary, July 15, 2010 (http://www.whitehouse.gov/the-press-office/remarks-president-holland-michigan-investing-clean-energy).

A123 Systems in Livonia, Dow Kokam in Midland, Fortu PowerCell in Muskegon, Johnson Controls-Saft in Holland, and Safti3 in Ann Arbor. So as you can see, this is a very exciting time in Michigan," Mr. Main said. "We are giving birth to an entire new industry in North America."

Mr. Main cited several people as driving forces in creating this symposium and bringing it to Michigan: Dr. Charles Wessner, director of technology, innovation, and entrepreneurship at the National Academy of Sciences; McAlister Clabaugh, program officer for the NAS Board on Science, Technology, and Economic Policy (STEP); and Dr. Mary Good, professor and founding dean of the College of the Donahue College of Engineering and Information Technology at the University of Arkansas at Little Rock and a STEP board member; and U.S. Senator Carl Levin.

Ranked by *Time* magazine as one of America's 10 best senators, "Senator Levin has fought tirelessly to strengthen and grow Michigan's economy," Mr. Main said. "He has been a long-time advocate of programs that provide for joint industry and government partnerships, including the development of advanced vehicle technology." These efforts led to the growth of the U.S. Army's National Automotive Center in Warren, Mich., he said, "which has played an important role in the development of advanced technologies for military use, often in conjunction with the private sector," he said.

Sen. Levin's efforts are helping assure that electric hybrid cars and trucks will have batteries stamped "Made in America" and "Made in Michigan," Mr. Main said. "You can get to know and admire government leaders, as I have over the years," he said. "But you will be hard-pressed to find one as energetic, as focused, as knowledgeable, and as generally committed to the work of building a better place to live, work, play, and raise one's family as our senior Senator."

Mr. Main welcomed Sen. Levin to the podium.

Opening Remarks I

Carl Levin
United States Senate

Senator Levin began by remarking that "this is really an opportunity for me to talk about one of my favorite subjects."

Senator Levin thanked the Michigan Economic Development Corp. for all of its efforts to help Michigan's economy. The work being done by the MEDC and attendees of the symposium "is enormously important to us, not just here in Michigan," he declared. "I am proud to be in the center of an industry that not only will burgeon, but also be important to our country, to our national security, and to the national economy as well. You are going to be making a huge difference in people's lives." Besides being interesting from a technical and marketing perspective, he said, the effort "is important to people who don't know an advanced lithium-ion battery from an Energizer bunny battery."

One reason to believe advanced batteries will "become a major industry" and usher in "major change in the way we move our people" is that President Obama, his Administration, and a majority of Congress "understand the transformative power of electric vehicles," Senator Levin said. "We also understand that we cannot unleash that power unless government partners with industry to make that happen."

Attitudes toward collaboration between government and industry have shifted dramatically in Washington, Senator Levin said. "A few years ago, anyone who suggested that government work closely with industry was accused of supporting an 'industrial policy.' If that industrial policy label stuck to anything, it was a kiss of death," he recalled. "That was not too long ago. That was a fact of political life."

Today, policymakers "understand that American companies are not only competing against foreign companies," Senator Levin said. "They are competing with countries and governments who support their domestic industries. We learned that. It took us too long. It put us at a competitive disadvantage to learn about the realistic necessity of that partnership." These days, "the question no longer is about whether government should be teaming up with industry," he said. "The question is about what we need to do, how we do it, and with what timeline."

Government and industry are "off to a great start," Senator Levin said. This is illustrated by the government's decision to invest more than $2 billion in advanced battery and related technologies, as well as in other areas to promote electric vehicles. "We need to make mass-market electric vehicles a reality, and that means a lot more than $2 billion for batteries," he said. "It also has meant a $5 billion investment aimed at electrifying—literally and metaphorically—the American transportation sector." The federal money is being combined with private investment, much of it by companies represented at the symposium in Livonia, he noted. "Factories are going up. Batteries are beginning to roll off assembly lines," Senator Levin said. "But that is just the beginning. Congress knows we have just begun this effort. The White House knows the same thing. That is true for research and development facilities and true for factories."

The next crucial step "is to figure out how to make electric vehicles affordable and sustainable in a country that has spent more than a century shackled to oil," Senator Levin said. "There is much left to do." Indeed, he said, now that the industry has turned the corner toward electric vehicles, "more challenges lay ahead of us than behind." He listed a number of major questions that must be answered:

- "How do we bring down the cost of these vehicles so that they are affordable not only to the greenest consumers, but also affordable to families who are stretching their paychecks to make ends meet?"
- "How do we give drivers who have spent their lives knowing that a gas station is right down the road confidence that the necessary infrastructure that is needed to recharge these new vehicles will be there for them?"
- "How do we give them batteries that they can depend on in the range that they will need?"
- "How do we produce enough electricity to keep these vehicles rolling, even during peak hours without adding even more harmful carbon into the atmosphere in the process?"

The common denominator in each of these challenges is "the need for the determination to see all these things through," Senator Levin said. For the power of electric batteries to be fully realized, policies aimed at promoting electric vehicles must have staying power. "We need that determination because we are beginning to fight a revolution, a revolution in transportation for the sake of our economy and the sake of our planet," he said. Policymakers must remain determined "not to allow oil-producing countries in the Middle East to throw us off our course of energy independence by slashing the cost of oil, as they have done in the past when we move toward breaking our addiction to their product," he said.

Based on current trends in technology development, consumer demand, and government incentives, Senator Levin noted, industry projections indicate that electrified vehicles can account for 10 percent of vehicles sold in the U.S. by 2020. But this bar should be raised. He urged industry experts to find an answer to a vital question. "What we need is for folks like you to tell us what it would take to double that goal," he said. "Tell us in Congress and tell us in the Administration what it would take to achieve a goal that would truly inspire the nation. What would it take to achieve a goal that is bold enough to move us beyond the incremental, miles-per-gallon-here-and-there battles on standards, and instead start us down a truly revolutionary path?" Industry experts should advise Washington on the kind of research support and regulatory support companies and academic laboratories need, he said. "Do we need new efforts to insure we have the raw materials?" he asked. "What investments are needed in new electricity generation? What would be needed to build the infrastructure to support that many electric vehicles? What would it take to sustain us once we are on that path?"

The National Academy of Sciences "will help guide us there," Senator Levin said. "But we will need the practical help, practical support, and practical advice of those that are in the industry."

A century and a half ago, young man working in Port Huron, Mich., as a telegraph operator "spent his every spare moment buried in science and technology books," Senator Levin noted. "His name was Thomas Edison. He combined that knowledge with his own remarkable vision and curiosity to begin a revolution that would literally light the world."

Now, America is "on threshold a second electrical revolution," Senator Levin declared. This revolution will "transform our streets, our economy, and the lives of millions of workers who could go to work producing vehicles that will dominate this century--just as Edison's light bulb dominated the last century." At the same time, this second electrical revolution can help restore America's manufacturing base, boost its economy, and save the planet from the fundamental threat of warming, he said. "You and folks like you can make this happen," Senator Levin said. "Tell us what you need to get us there, and I can commit to you that most of my colleagues and I in the Congress will do everything that we can to give you the tools and support you need to make this a great nation."

Senator Levin thanked the attendees for their efforts and for "helping to insure the younger generation will join in this truly peaceful revolution to make this world a far better place."

Opening Remarks II

INTRODUCTION

Charles W. Wessner
The National Academies

Dr. Wessner said he was inspired by Senator Levin's speech and encouraged by his commitment that there will be consistent Congressional support for the advanced-battery initiative. The U.S. system, however, requires not only legislature support but also "an executive who is willing to execute."

Therefore, Dr. Wessner said he was pleased to introduce Dr. Sridhar Kota, assistant director for advanced manufacturing at the White House Office of Science and Technology Policy. Dr. Wessner noted that Dr. Kota has assumed his Washington responsibilities while on leave from the University of Michigan, where he is a professor of mechanical engineering, adding that Dr. Kota is "ideally situated for our discussions today because he is also an entrepreneur," having founded Flexis Inc., which develops fuel-efficient adaptive aircraft wings and high-efficiency wind turbine blades.[2]

[2] Dr. Sridhar Kota founded FlexSys Inc. in 2000 to develop and commercialize his patented design of a shape-morphing adaptive control surface of an airfoil. Dr. Kota is a pioneer of the bio-inspired concept of distributed compliance, which allows structures to adapt their performance in response to different conditions. The technology is being applied in aerospace, automotive, and other industries.

Dr. Kota received the American Society of Mechanical Engineers (ASME) Machine Design Award, which is the highest award dedicated to engineering design, as well as the ASME Leonardo DiVinci Award. "We are very fortunate that we have an exceptional individual like Dr. Kota, both at the White House and here to focus on the needs of this state and this nation."

Sridhar Kota
White House Office of Science and Technology Policy

The United States still leads the world in manufacturing, accounting for some 21 percent of global production, Dr. Kota noted. However, that leadership has eroded due to "the decisions made decades ago for off-shoring and outsourcing." The erosion also can be seen in indicators such as the rising U.S. trade deficit in high-tech goods with nations such as China. That deficit has been growing since 2001.

When it comes to understanding the importance of U.S. manufacturing, the Obama Administration "gets it," Dr. Kota said, "not only for economic security but also certainly for energy security and national security." Therefore, the Administration has made considerable investments in basic research as well as what he referred to as the "manufacturing commons."

The product development cycle in the U.S. generally begins with basic research, where the federal government heavily concentrates its investment, Dr. Kota explained. This research leads to discoveries and inventions at universities and national laboratories. But although the United States still leads the world in basic research, "that is just a first step," he said. "We need to go beyond that." Not every discovery will lead to a commercial product, as you all know well." Therefore, the U.S. must invest in translational research to develop proofs of concept, see which pan out, and "take it further to go to manufacturing."

To move to large-scale production, investment in the "manufacturing commons" are needed, Dr. Kota said. This commons includes engineering R&D, skills, components infrastructure, equipment, and standards. Platform technologies must be developed. By building the manufacturing commons, he said, "you build new products and innovations--and the cycle continues."

Manufacturing is very closely tied to innovation, Dr. Kota said. "If you don't have those manufacturing commons in place, we are not going to be able to innovate next-generation products. There is no doubt about that." The Administration, therefore, is increasing spending on R&D but also is investing in infrastructure and developing innovation policy for "creating new industries and growing and sustaining existing industries," he said.

The essential requirements for establishing new industries are radical technological innovation, early adaptation of breakthrough technologies, and access to capital, Dr. Kota noted.

In terms of innovation, the Administration is trying to fill a gap in the process. Most federal investment goes into scientific discovery, Dr. Kota explained. Twenty or 30 years ago, large corporate research labs such as ATT's Bell Labs and Xerox Park took basic research and turned it into technologies and applications. Since those big labs closed, most corporations concentrate their R&D spending on commercial development, "leaving a gap in the middle" of the process, he said. To fill that innovation gap and establish the manufacturing commons in the battery industry, the Administration is encouraging public-private partnerships.

The federal government also is trying to promote early adoption of batteries to enable manufacturers to scale up, Dr. Kota explained. The U.S. Postal Service and the Department of Defense both are buying more electrified vehicles, for example. To improve access to capital, the federal government has established a number of policies and programs. They include DOD loan guarantees and new Export-Import Bank programs.

To grow and sustain existing industries, technological and business innovation is needed, of course, Dr. Kota said. But industries also need fundamental infrastructure. That includes "a skilled workforce at all levels, not only trained manufacturing workers for next-generation emerging technologies but also college graduates with multidisciplinary skills to deal with the next-generation, innovative energy engineers," he said. New programs also are being considered to provide tools that can improve quality, time, and cost, such as computer modeling and simulation, he said. American productivity remains high, Dr. Kota pointed out. "But it has been coming down steadily," he said.

Many challenges facing U.S. production are addressed in a December 2009 document issued by the White House called "A Framework for Revitalizing American Manufacturing." The framework identifies key challenges and drivers of global competitiveness in manufacturing and seven principles to strengthen the U.S. manufacturing base. They are:

- Double R&D budgets for key science agencies.
- Improve coordination of manufacturing-related R&D.
- Explore new options to stimulate innovations and technological breakthroughs.
- Make the research & experimentation tax credit permanent.
- Spur innovation in manufacturing by expanding the Technology Innovation Program (TIP).
- Pursue structural reforms that support innovation and production.
- Protect intellectual property rights.[3]

The Administration is acting on each of these priorities in the 2011 federal budget, Dr. Kota explained. In terms of technology, it boosted funding for the National Science Foundation to advance basic research, for example, and called for $150 million by 2015 for the Technology Innovation Program at the National

---

[3] Executive Office of the President, "A Framework for Revitalizing American Manufacturing," op. cit.

Institute of Standards and Technology. The TIP program, Dr. Kota noted, "was largely neglected in the past few years but now is being strengthened."

The budget also calls for investing $12 million in university innovation centers that focus on developing proofs of concept and prototypes, an additional $10 million for nano-manufacturing, and $300 million for the Advanced Research Projects Agency-Energy, or ARPA-E, a new organization within the Department of Energy that promotes R&D in new energy technologies. Other efforts under the Department of Commerce and the Office and Science and Technology Policy promote R&D commercialization at universities.

Business-related investments by the Administration include access to capital under the DoE's 1703 and 1705 loan guarantee programs,[4] greater access to capital for exports, the 1603 program that gives cash grants in lieu of tax credits,[5] the 48 C manufacturing tax credit,[6] and the advanced vehicle manufacturing loan program. The manufacturing tax credit, Dr. Kota noted, originally provided $2 billion that leveraged more than $7 billion in private investment. But the program proved so popular that the Administration expanded it by another $5 billion, he said.

The advanced-battery program exemplifies the new manufacturing strategy. "Battery technology, some may say, was a good example of something that was largely invented here and manufactured elsewhere," Dr. Kota said. "But now, the stars are all aligned. I think there is great opportunity right now for us to regain leadership in battery technology." Not only is the federal government making investments, but there also is great support by states such as Michigan, local government, and public-private partnerships, he said. "Investing in R&D and the manufacturing commons will give us the infrastructure we need," Dr. Kota said.

Half of the $2.4 billion in Recovery Act funding for the battery industry went to Michigan, Dr. Kota noted. Some $4.5 billion in federal investment in smart-grid technologies also will help manufacturing of advanced batteries, as well as devices such as smart meters, he added.

Another example of federal commitment to advanced batteries, Dr. Kota noted, is a July 21, 2010, memo by Office of Management and Budget Director

---

[4] Section 1703 of Title XVII of the Energy Policy Act of 2005 ("EP Act 2005") authorizes the DOE to issue loan guarantees to acceleration commercialization of technologies that "avoid, reduce, or sequester air pollutants or anthropogenic emission of greenhouse gases." Section 1705 of the EP Act is a temporary program set up under the American Recovery and Reinvestment Act authorizing the DOE to make loan guarantees to renewable energy systems, electric transmission systems and leading-edge bio-fuels projects that commence construction no later than Sept. 30, 2011.

[5] Section 1603 of the American Recovery and Reinvestment Act created a program administered by the U.S. Department of Treasury that extends grants covering between 10 percent and 30 percent of the cost of certain renewable-energy property.

[6] The Advanced Energy Manufacturing Tax Credit was authorized in Section 1302 of the American Recovery and Reinvestment Act and also is known as Section 48C of the Internal Revenue Code. It authorizes the Department of Treasury to award $2.3 billion in tax credits to cover 30 percent of investments in advanced energy projects, to support new, expanded, or re-equipped domestic manufacturing facilities.

Peter Orszag and John Holdren, director of the Office of Science and Technology Policy. The memo identified six science and technology priorities. One was: "Prioritize R&D on advanced vehicle technologies, particularly modeling and simulation of lightweight materials and their manufacturing processes, batteries, and hybrid power trains; and systems integration and demonstration of advanced vehicle platforms."[7]

Federal investments targeting advanced batteries include the DoE's Energy's Vehicle Technology Program,[8] ARPA-E, and the Department of Defense's program to electrify its fleet of non-combat, tactical, and combat vehicles. The departments of Defense and Energy are collaborating on another program to electrify the U.S. Army's fleet. The Army has committed to cutting its fuel consumption by 20 percent in the next 10 to 15 years, Dr. Kota noted. The electrification of the Army's fleet "gives us an opportunity as an early adopter and an opportunity for you guys to take your ideas to the Army to use as a proving ground." The DoE/DOE collaboration involves great technologies coming out of the DoE, whether they are batteries or composites."

The fact that the Army's Tank-Automotive Command Research, Development, and Engineering (TARDEC) headquarters is in Michigan "is a great opportunity" for DoE-DOE collaborations to commercialize component technologies and integrate them into vehicles, Dr. Kota said. Rather than "just present a shiny object," such collaborations can "prove the cost and manufacturing feasibility for next-generation, advanced fuel-efficient vehicles," he said. "I think that is great."

Michigan's investments in the industry are paying off. "Michigan is well poised to become the battery manufacturing capital, thanks to efforts by the MEDC, Governor Granholm, and the state's Congressional delegation, who have made tireless efforts and unwavering commitment to this industry," Dr. Kota said. The fruits of these efforts include big investments by Johnson Controls, A123, and Dow-Kokam. "There also are very many start-ups coming out as a result of investments both by the federal and statement government," he noted.

Great research is coming out of America's national laboratories and universities, but now comes "the hard work of translating that into real-world, practical solutions," Dr. Kota said. Major hurdles remain for the advanced battery industry, such as cost, performance, safety, battery life, and

---

[7] From M-1-30 Memorandum for the Heads of Executive Departments and Agencies, by Peter R. Orszag, director of Office and Management and Budget, and John P Holdren, director of Office of Science Technology Policy, "Science and Technology Priorities for FY 2012 Budget," Executive Office of the President, July 21, 2010 (http://webcache.googleusercontent.com/search?q=cache:vLUw6QLwTosJ:www.whitehouse.gov/omb/asset.aspx percent3FAssetID percent3D2852+orszag+holdren+july+21+memo+science+priorities&cd=1&hl=en&ct=clnk&gl=us&client=firefox-a)

[8] The Vehicle Technologies Program is administered by the Energy Efficiency and Renewable Energy Office of the Department of Energy. It funds projects aimed at developing "leap frog" technologies that will lead to more energy-efficient and environmentally friendly transportation.

manufacturing. "Each of those is a great, tough challenge. And at the end of the day, you have to meet all of them," Dr. Kota said. Just meeting cost requirements but not solving performance problems won't be sufficient, for example. "Meeting just one of them is hard," he said. "Meeting all of them is a daunting task."

Many experts and programs are working to meet those challenges, Dr. Kota pointed out. For instance, ARPA-E is working on "potential breakthroughs in new battery chemistries that are two or three or five times better than current technologies," he said. It also is working on new manufacturing methods "that could change the game altogether." In addition to components, important R&D efforts are underway in composites for more fuel-efficient vehicles, he said.

At a higher level, public-private partnerships through the DOE and DOD are collaborating on system integration for electric vehicles, Dr. Kota pointed out. U.S. efforts to promote a smart power grid "tie in very well with the efforts in electric vehicles and hybrids," he said. "So there are great opportunities for very talented people to meet these challenges."

To conclude, Dr. Kota posed several questions. The first regards electric vehicles. "What technological breakthroughs are needed for a sustainable plug-in hybrid and electric vehicles industry?" he asked. "You are the experts. If you could identify those that would help guide the things we do in Washington."

The second question is about batteries. "What kinds of partnerships and business models and policies do we need to gain global manufacturing leadership in these technologies?" he asked. Washington's role is another. "We have many programs and policies in place," Dr. Kota said. "We want to hear from you what is working and not in terms of how the federal government can be more effective in strengthening the battery industry." The fourth question regards economic growth. "As President Obama said, this is not about creating a battery program," he said. "It is about unleashing private-sector growth. The people rights here are testament to that private-sector investment and partnership."

Michigan has more than 100 years of experience in product development, manufacturing, and engineering, Dr. Kota said. "There is no better place in the world to try to create a vibrant and globally competitive battery industry."

Opening Remarks III

Jennifer Granholm
State of Michigan

Governor Granholm welcomed the participants and welcomed them to Michigan for "this conference on the achievements, challenges, and opportunities in developing a globally competitive battery industry." She said the organizers "sure picked an appropriate time—and you better believe you picked an appropriate place for this conference."

Governor Granholm noted that July 14, 2010—just one week prior to this conference—was the anniversary of the DOE announcement that it would invest $2.4 billion to support advanced battery and electric vehicle manufacturing development. Michigan "is well underway to becoming the advanced battery capital of the world," she said.

Vice President Joe Biden came to Michigan in August 2009 to announce that $1.35 billion of those grants would go to 12 projects in Michigan, Governor Granholm noted. "That total was more than all of the other 49 states combined," she said. Michigan succeeded because it targeted advanced batteries several years earlier as "a sector we wanted to grow to diversify Michigan's economy and to create jobs," she said.

Michigan's tax credits for advanced battery projects "were the first in the nation and, by the way, the most aggressive in the nation," she said. "When I signed those battery credits into law, it sent a clear signal that Michigan is very serious about being a leader in this industry."

As a result of those credits and the DOE grants, "a whole advanced battery supply chain is taking root from the Detroit area to the shores of Lake Michigan," Governor Granholm said. Michigan now has 16 advanced-battery projects underway, representing around $6 billion in capital investment. Those plants are expected to create some 62,000 new jobs over the next 10 years, she said.

That supply chain includes anodes, cathodes, separators, and electrolytes, as well as companies building cells, integrating them into battery packs, and integrating them into electric vehicles, Governor Granholm explained. "The whole spectrum is right here in Michigan," she said. "This is an exciting time as we begin moving in earnest to a clean-energy economy in the United States."

The emerging advanced-battery industry is the result of collaborative partnerships between the federal government, state and local governments, and the private sector, Governor Granholm said. "It is critical that these partnerships continue and that they grow stronger," she said.

Congressional help also is needed, she said. That includes passage of legislation to expand the advanced energy manufacturing tax credit and that continues federal tax credits to consumers buying electric vehicles, "at least until the cost of manufacturing lithium-ion batteries is comparable to that for internal combustion engines," Governor Granholm said.

A clean-energy economy not only will "create millions of jobs but also will reduce our dependence on foreign oil and enhance our national security, our energy security," Governor Granholm said. "It is a win, win, win for our nation."

Manufacturing the "key components for the clean-energy economy" in the United States is "absolutely critical," she said. "The batteries, the wind turbines, the solar panels—right here. And Michigan, I just want to let you know, intends to lead the way in clean-energy manufacturing. We have manufacturing in our DNA." Michigan has manufacturing capacity, a workforce, and "engineers who know how to solve problems," Governor Granholm said. "So we are very excited. We are very bullish on the opportunities that clean-energy manufacturing provides."

Governor Granholm thanked the National Academies' STEP Board "for honoring our progress in the advanced battery industry in Michigan by convening this conference right here in our state." She said that "I hope you have a fantastic conference, I hope you enjoy your stay here, and I invite you all to visit again to experience everything that is pure Michigan."

Overview of NAS Study: Building the Battery Industry for Electric Vehicles

Mary Good
University of Arkansas at Little Rock

Dr. Good said she was pleased Senator Carl Levin could attend the symposium. "He has certainly been a big help and big supporter of this sort of activity for a very long time," she said. "When we were seeing some of these things off the ground a long time ago, he was a supporter. It is pleasant to see some of these things begin to come to fruition."

The strong attendance at the symposium in Livonia reflects a "high degree of interest in what has been a remarkable effort to build a U.S. battery industry," said Dr. Good, a member of the National Academies' Board on Science, Technology, and Economic Policy. The Department of Energy did an excellent job with the battery program by "selecting the winners and moving the funds in a very short time frame," she said.

The leadership of Patrick Davis and his colleagues at the DOE's Vehicle Technologies Program "is to be applauded," Dr. Good said. "I don't think many people really understand the chore that was given to the Department of Energy people to take this stimulus money and effectively get it out into the community to use in a very short period of time." Accomplishing that isn't easy in the private sector, which can move without constraints, she noted. "But doing it in a federal agency is nearly miraculous, in my opinion, and they have done a superior job."

The U.S. is facing "intense and growing competition from other nations," Dr. Good noted. "As many in this room know, we are competing not only against other companies but also other nations and regions around the world for the well-paid jobs and improved living standards that come from the leadership and development derived from manufacturing new technologies and new products."

The White House report on manufacturing is very encouraging, Dr. Good said. "We really have to get the leadership of the country to understand we cannot abandon manufacturing in the United States." Without manufacturing, the U.S. would be lucky to even hold on to an acceptable standard of living, she said. "This continued leadership is essential if we would like our children and grandchildren to at least have something close to the same standard of living we have."

Michigan "understands the intensity and global nature of this challenge," Dr. Good said. The state has "shown great leadership and taken concrete steps to face this competitive challenge and is to be congratulated," she said. Given Michigan's economic and employment problems, the state's leadership "is close to miraculous."

The White House also is committed to securing leadership in energy and transportation technologies, Dr. Good said. "We are really fortunate and happy to have people like Dr. Kota, who on is taking time off from his Michigan assignments to be part of this discussion in the White House." Some "really good people" have been appointed to address such issues in the White House, she noted, "and he is a really good example of the kinds of folks that are now available in the Administration to make these kinds of things work."

The symposium's goal is to highlight the challenges and opportunities for Congress, the DoE, Michigan, and other states as they work to develop an advanced battery industry in the United States, Dr. Good explained. "And it is indeed a challenge," she said. "To start a new industry like this and be competitive is not a simple thing to do."

Above all, the purpose of the conference is to seek expert opinion "on what is working and what is not working," she said. The conference will not produce conclusions on what the federal government should do, she said. "What we want to do is provide information and assessments that help people working on these problems to make good judgments. To do that, they really need all the good input they can get."

The symposium is part of a broader effort by the National Academies to study selective state and regional programs, Dr. Good explained. "This particular committee's aim is to try to identify the best practices with regards to their goals, structures, instruments, and modes of operation," she said. It also is studying best practices regarding fund levels and mechanisms, as well as "the challenges, accomplishments, and evaluation efforts of these programs."

The STEP board also is studying how regions are capitalizing on state and federal investments in "developing a knowledge-based, innovation-led economy," Dr. Good said. Many economic development efforts around the nation are being led by state governments, she noted.

Dr. Good expressed the STEP board's gratitude for the support and insights of the MEDC and DOE for "bringing together the battery community in this room and their support of this event." She especially thanked Gary Krause of the MEDC, who has been instrumental in making the conference happen. McAlister Clabaugh of the National Academies and David Howell and Jim Miller of the DOE also were instrumental. Dr. Good also thanked A123 Systems and the Michigan University Research Corridor for supporting the conference.

The National Academies will follow with another conference on advanced batteries in Washington, D. C., Dr. Good noted. At that event, "we will expand on the issues raised" at this symposium. The board decided to come to Michigan for the initial conference "because this is where the industry is and where the federal and state governments are making big commitments in resources," she said. "So we wanted to hear first-hand how the industry is doing."

There are many policy issues relating to this new industry, Dr. Good explained. Experts at the conference will describe the state of the industry and highlight areas of further attention for R&D, manufacturing, the network of suppliers, and the type of workforce needed to keep the industry competitive, she said. Speakers also will discuss how to expand the market for electric vehicles and "hasten the widespread use of advanced batteries," she said. Representatives from state and federal agencies will explain programs providing R&D and manufacturing support to the battery industry.

Dr. Good said Dr. Charles Wessner, director of the National Academy of Science's Program in Technology, Innovation, and Entrepreneurship, "has really been a major spark plug." She congratulated him and his staff for organizing the program. "To get these kinds of programs off and running and in good shape in the time frame of this one is rare for the Academy," she said.

Keynote Address

INTRODUCTION

John R. Chalifoux
Original Equipment Suppliers Association

U.S. Senator Debbie Stabenow (D-MI) "has been a true friend and fighter for Michigan's emerging clean-energy industry," said Mr. Chalifoux, the OESA's marketing and business development vice-president. She is a "nationally recognized leader" who has "the ability to organize coalitions to get things done for Michigan and for our nation."

Senator Stabenow serves on the Senate Energy and Natural Resources, Finance, Agriculture, and Budget committees. This gives her "a unique role in shaping our nation's manufacturing, trade, and energy policies," Mr. Chalifoux said. "Senator Stabenow is bringing manufacturing issues to the forefront in Congress and is committed to making sure our nation has a 21st century manufacturing strategy to cultivate America's leadership in alternative energy."

The senator is co-author of the Energy Manufacturing Tax Credit and of consumer tax credits for the purchase of electric and hybrid vehicles, Mr. Chalifoux noted. Senator Stabenow also championed passage of the advanced battery manufacturing grants and retooling loans for advanced vehicle production, he added. She authored the Senate version of the "cash for clunkers" program that gave U.S. car-owners financial incentives to replace old cars with more fuel-efficient ones. This program is credited with creating 60,000 jobs in 2009, he noted. Senator Stabenow's leadership led President Obama to appoint her to the President's Export Council, created to advise the President on export issues and boost exports by American companies, he said.

Mr. Chalifoux said he "would be remiss as a member of a supplier's trade association" if he didn't add that Senator Stabenow has led the push for legislation to give R&D assistance to U.S. suppliers. The bill, S. 2843, the Advanced Vehicle Technology Act of 2010, "will help vehicle manufacturers and suppliers develop and implement technology for more fuel-efficient vehicles and components," he said.

The legislation also will make significant changes to the Advanced Technology Vehicle Incentive program managed by the DoE, Mr. Chalifoux explained. Senator Stabenow and U.S. Sen. Ron Wyden (D-Ore.) introduced legislation to extend the existing program to medium- and heavy-duty vehicle manufacturers and suppliers and "improve the program to allow for greater supplier participation overall," he said. That provision passed the Senate Energy Committee in early July 2010.

The OESA supports and "greatly appreciates the Senator's work on behalf of suppliers," Mr. Chalifoux said.

Debbie Stabenow
United States Senate

Sen. Stabenow thanked the National Academies for "pulling this session together" and recognized Dr. Wessner for his leadership. She also acknowledged the efforts of TARDEC "and the important efforts of leaders around this room."

Clean-energy policy in the U.S. is fundamentally about jobs, Sen. Stabenow said. "It is about other things. But when you come from Michigan, it is all about jobs," she said. As a member of the Senate Finance, Energy, and Agriculture committees, "I am laser-focused on all of the things you are talking about." With the leadership of the Michigan Economic Development Corporation and Governor Granholm, "we have very much been focused on clean energy," she said.

Although she said she is not eager to go back to Washington's hot summer weather, Sen. Stabenow said she is returning "because we are going to continue to focus on jobs and an important small-business bill that will help suppliers and create capital for small businesses."

The topic of this conference is "incredibly important," not only for Michigan but for the country, Sen. Stabenow said. She noted that a man named Steve Pernell, who had worked on the assembly building cars for GM, now is building test models for the Chevy Volt. Mr. Pernell told a Fox News reporter that he feels the pressure to build the car well because "the success of the Volt is a matter of do or die."

Mr. Pernell is right, "and not just for GM," Sen. Stabenow said. "Building the next generation of energy-efficient vehicles is do-or-die for all of the automakers, for the state of Michigan, and for America." The success or failure of these vehicles will largely depend on the quality of the batteries that will power them, she said. "So this is a very, very important discussion and an important effort that we all need to be continually engaged in."

There has been "incredible spending" by Asian nations on battery technology, Sen. Stabenow pointed out. "Japan, Korea, and others have taken the early lead," she said. "China has now gotten into the game big time. China is now spending about $288 million a day to beat us on clean energy. So this is a race."

American companies have been competing against countries for years, Sen. Stabenow stressed. "Finally our country is beginning to get in the game in terms of partnership with our businesses," she said. "It is incredibly important that we ramp this us as fast as we can." The last thing the U.S. needs to do "is go from a dependence on foreign oil to a dependence on foreign technology," she said. "And if we don't continue to push, that is what is going to happen."

The U.S. is starting to turn the corner, Sen. Stabenow said. "There are many positive things happening," she said. "But we are starting from a position of being behind on this." The strategic focus by U.S. companies, the federal government, and states "will get us there," she said. "We are going to have to be serious and we are going to have to be focused."

The U.S. now has an Administration "that understands the importance of making things in this country," Sen. Stabenow said. For more than a decade, the only thing that mattered was cheap prices. Americans did not care where products were made, she said. "We have been losing our middle class as a result of that," she said. "It matters where things are made. I happen to believe that we can only have a strong economy and middle class if we make things and grow things and add value."

Government in the U.S. now is working on this, Sen. Stabenow said. We are seeing a different discussion now." As a member of the President's Export Council, Sen. Stabenow said she is "enthused" about the efforts to double U.S. exports in the next five years. "We want to export our products, not just our jobs," she said. "I think that is the focus for us—jobs and innovation here."

The U.S. still should form global partnerships, Sen. Stabenow said. "We are in a global economy, so of course we partner," she said. "But we should not take our eye off the fact that we want our jobs here."

The Recovery Act was really about starting to make the investments needed to bolster U.S. industry, Sen. Stabenow said. Michigan has been a beneficiary, perhaps more so than any other state. She noted that more than $12 billion was invested in advanced vehicle technology in the previous 18 months through the Recovery Act.

About $2.3 billion of that went exclusively to batteries. "I am very proud of the fact that Michigan has received over 50 percent of that money," she said. "I told the President and Vice President directly that we like that ratio. We would like it with every program." The Administration has recognized Michigan as the center of advanced battery manufacturing, "and that is great for us in the long run," she said.

Sen. Stabenow noted that President Obama was in Holland, Mich., recently to celebrate the opening of Compact Power's facility. Between the Advanced Manufacturing Tax Credit and the efforts around batteries, she noted, some 2,000 jobs are expected to be created in the Holland area over the next few years. The President also visited Chrysler and GM facilities in Wayne County to explain the importance of the U.S. auto industry and the impact of U.S. investments.

The investments are important not only for OEMs, Sen. Stabenow said. They also are important for suppliers. "If we did not have an American automobile manufacturing industry, the ripple effect across all of our suppliers and all of our industries would be absolutely devastating."

A123 is a "great story" about bringing business back from Asia, Sen. Stabenow said. The company is opening a new plant in Livonia, near the site of the symposium, she noted. "Creating that new technology and jobs is important for us," she said. There are nine new battery plants under construction or operating "from one end of the state to the other," she said. They include Johnson Controls in Holland, Dow Kokam in Midland, and facilities by Ford, GM, Magna, and Chrysler.

It also is important that battery investments have been made with universities, Sen. Stabenow said. The University of Michigan, Michigan State, and Michigan Tech all received battery grant funding for R&D and partnerships with business, she explained. "Our great universities are such an important part of where we want to go," she said. "We have more engineers and high-quality workers than any other place in the country. When we talk about clean energy, manufacturing, and developing new technologies for our country, I know that means we are going to do well in Michigan because we have the talent here." One example of this talent is Ann Marie Sastry, she noted, who founded Sakti3 Inc. at the University of Michigan "to translate groundbreaking research to manufacturing advanced batteries and is putting that to work right here in Michigan."

The Advanced Manufacturing Tax Credit program also has boosted manufacturing in Michigan. "For the first time, we have put into law that if you are buying equipment or investing in a plant for clean-energy manufacturing, there is a 30 percent tax credit," Sen. Stabenow explained. "That is important, and is part of what can make us competitive." Twelve Michigan companies have received the tax credits, which have gone to 46 states. There were three times as many qualified applicants for the tax credits than there was available funding, she said. As a result, President Obama and Vice President Biden asked Congress for another $5 billion. "I was able to put that into the budget resolution, and I hope we will get that done by the end of the year," she said. "A lot more businesses and manufacturers are waiting to receive help from this."

The Senate Energy Committee also is doing important work on electric vehicles, Sen. Stabenow said. In July, she joined U.S. Rep. Gary Peters of Oakland County to include suppliers and medium- and heavy-duty vehicles to an advanced-vehicle technology program. A loan program for retooling also was expanded to include medium- and heavy-duty vehicle suppliers.

Another piece of legislation coming out of the Energy Committee is the Promoting Electric Vehicles Act.[9] Sponsored by Senators Byron Dorgan, Lamar Alexander, and Stabenow, the bipartisan bill calls for the DOE to make concerted efforts to help 15 communities develop infrastructure needed for electric vehicles. She said the program will help set up charging stations and address "whatever the process is" to have charging at home. "I'm starting to hear that it can take months to get a permit and that folks in local government

[9] The Promoting Electric Vehicles Act of 2010 (S. 3495) sponsored by Sen. Byron Dorgan (D-ND), Sen. Debbie Stabenow (D-MI), and Sen. Lamar Alexander (R-TN) calls for providing incentive programs to create "deployment communities" across the U.S. stations for purchasing electric vehicles and set up charging facilities. The Senate Energy and Natural Resources Committee approved the bill on July 27, 2010.

are trying to figure out how to make this all work in terms of infrastructure for electric vehicles," Sen. Stabenow said. "We would like very much to pick a number of communities on a competitive basis and try to really jump-start start that. We want to create models of how to develop that infrastructure as quickly as possible."

It will not be enough just to have electric vehicles on the road, Sen. Stabenow said. "They have to be consumer-friendly," she said. "And that is a real challenge for us." Reducing battery costs by 70 percent by 2015 is important, she said. "But again, we have to make sure that the infrastructure is there as well." Success in electric vehicles "depends on a whole range of things," Sen. Stabenow said, "from smart-grid technologies to the permitting process so that you can make sure you have what you need in your home."

As with any new technology, the prices of electric cars will be high until production volumes increase, Sen. Stabenow pointed out. The federal government can help build the market as a buyer. The government will buy the first 100 Chevy Volts, for example. "I would like to add a few zeros to that," she quipped. "We need to be doing the same with Ford and Chrysler. The federal government can and should be a purchaser." She noted that the federal government has some 700,000 vehicles, including those operated by the U.S. Postal Service and the military. A new Senate bill encourages federal agencies to purchase electric vehicles.

The government also has helped with consumer tax credits. Buyers of plug-in vehicles can get a $7,500 credit, Sen. Stabenow explained. Currently, these credits will go only to the first 2,500 purchasers of plug-ins until 2014. "I know if it works we will be expanding it," she said.

Sen. Stabenow said she is working on legislation to allow those tax credits to be given at the time a car is purchased. The experience of the Cash for Clunkers program, which she said "was amazingly successful beyond my wildest dreams," showed that it helps to allow people deduct a tax credit immediately off the price of car. "That is more helpful than waiting until you fill out your taxes the next year," she said. "It will create a lot more opportunity for people to be able to use that tax credit."

Sen. Stabenow said she is working on legislation to allow these credits to be applied at dealerships and to extend the credits to medium- and heavy-duty vehicles as well as for fleet purchases by businesses that buy 50 or more hybrids or electric vehicles. "There is a lot of interest right now," she said. "We have heard from a lot of fleet owners interested in buying electric vehicles." She added that she has seen the batteries being installed in trucks in Michigan. "There is no reason why we should be getting energy efficiency only from small vehicles," she said.

Fair trade is another important issue, Sen. Stabenow said. China has policies not only to invest aggressively, "which they certainly have every right to do in their own technology," she said. But China also has "indigenous innovation" policies that "are blocking our companies from the ability to sell to their government," she said. There also are requirements on transferring technology, evidence of currency manipulation, and problems with intellectual property rights. "There are a whole range of issues that involve creating a level playing field that we in the federal government have got to pay attention to and do something about," Sen. Stabenow said.

Senators Stabenow, Lindsay Graham, Jarrod Brown, and Russ Feingold have introduced the China Fair Trade Act of 2010[10] in response to Chinese policies directing the government to only do business with Chinese companies, she noted. "Basically what we are saying is that China, which has been part of the World Trade Organization 10 years now, agreed to abide by international law," she said. China has not signed the WTO's government-procurement agreement that allows companies from other nations to bid on public contracts, she pointed out. "Now they are moving much more aggressively," Sen. Stabenow said.

No longer does the Chinese government just require foreign companies to form Chinese partnerships. She recalled visiting China in 1995, when General Motors launched its first joint-venture. "It's now beyond joint ventures," she said. "It has to be a Chinese patent and it has to be a Chinese company. It is going way beyond what has been viewed as a level playing field in the past."

The new bill is aimed at simply saying, "If they are not going to let us sell to them, then we aren't going to let them sell to our government and use federal tax dollars to buy their goods until they open up their markets," Sen. Stabenow said. She said she disagrees with people who argue that fighting for fair trade in a global economy is about protectionism. "This is not about walls," she said. "It is about creating opportunities for our companies to be able to compete successfully and not have markets closed unfairly." It is great that the U.S. is partnering with other countries in industries like clean energy," she said. "But we have to have access to markets if we are going to meet our exporting goals."

The U.S. also needs a comprehensive energy strategy, Sen. Stabenow said. "We have to fully decide that we are in it," she said. "I believe that this is a critical part of our next economic wave." The U.S. should raise the price on carbon pollution and plow the resources that are raised into technology. "We can do that in a way that is a winner for us," she said. "But we have to make sure that we have certainty in the marketplace for everybody who wants to come in and invest in batteries and clean energy." Investors need certainty on policies and tax laws, she said.

The U.S. also has to make sure it creates capital for the front end. Sen. Stabenow noted that the Energy Bill now in Congress creates a mechanism to finance the first deployment of technology. "Everyone will invest in the third,

[10] The China Fair Trade Act of 2010 (S. 3505) was introduced on June 17, 2010, by Sen. Lindsay Graham (R-S.C.), Sen. Debbie Stabenow (D-MI), Sen. Russ Feingold (D-MN), and Sen. Sherrod Brown (D-OH). It would bar the U.S. government from purchasing Chinese products until China agrees to the Agreement on Government Procurement of the World Trade Organization.

fourth, fifth or tenth" product, she said. "But who is going to commercialize the first one?"

Other countries will make such investments, she said. "They will offer to build the plant and to give you the financing," Sen. Stabenow said. "We ought to have a way." She noted that bills in the House and Senate have proposed setting up a Clean Energy Development Administration that would help fund early-stage commercialization of new technologies. "We are very hopeful that we are going to be able to get this going," she said.

Sen. Stabenow explained that she and Sen. Graham of South Carolina chair a bipartisan effort on U.S. manufacturing. "I do believe we are seeing a real change in terms of how we focus not only on clean energy and batteries and technology, but also on manufacturing," she said. "I don't want to see us going through with batteries and electrification of vehicles what we did with the technology that created the iPod, where we have the President going to England, visiting the Queen, and giving her a great product of American ingenuity--an iPod made in China. Shame on us if we let that happen in batteries, or wind, or solar, or anything else in clean energy. Shame on us."

The U.S. still has a chance to succeed in the electric-vehicle and battery industries, Sen. Stabenow said. "They haven't left us yet. We are in a fierce race, but this is ours to capture," she said. "And my focus is to make sure that when those batteries and vehicles come off the line and the new technologies are being produced, they all say 'Made in America' again. That's when we win."

Sen. Stabenow asked those in the room to help generate "a sense of urgency about this." She said she has "talked to too many businesses who have a great idea, a great technology, or a great innovation who can't get the financing right now." Others are held back by factors such as certainty about tax policy. "I talk to venture capitalists all the time who are sitting on the sidelines saying they are waiting to see where energy strategy is going to be before they know where to invest," she said.

Help from those attending the symposium is needed "to kick this into gear," Sen. Stabenow said. "You folks are on the front lines. You know the facts. You know the reality of what is happening." Every single member of Congress must be told to stop thinking in terms of who is a Republic and who is a Democrat. "We can't afford that," she said.

Sen. Stabenow recalled that she was in a meeting with President Obama and backers of a bipartisan energy bill. "We were saying, 'Well, we don't know if we can get enough Republican votes or if we can get over the filibuster,'" she said. "I said, 'You know, while we are talking about this, China is cleaning our clock.'"

"This should be about the United States versus China," Sen. Stabenow said. "Not whatever else is going on in Washington. We need the help of the Academies and from businesses. Let's talk about getting the policies right. We can have differences on policy, but we are all on the same side here in this country."

Sen. Stabenow said the U.S. has "absolutely great opportunities" if it uses all of its intellectual capacity, talent, and resources. "There is no reason that we can't have the President give the Queen a wonderful new electric vehicle with all of the components, including the battery and cell, made in America."

Sen. Stabenow noted that Henry Ford once said that "what is right about America is that, although we have a mess of problems, we have great capacity—intellect and resources—to do something about them."

That is where America is right now, she said. "We may have a lot of challenges," Sen. Stabenow said. "But we have great intellect and great resources. Now is the time to put it all to work."

Panel I

The Federal Outlook for the U.S. Battery Industry

Moderator:
Charles W. Wessner
The National Academies

This panel "will be interesting" because it features "top-flight people who are intimately involved" with the issues of advanced batteries, said Dr. Wessner

He then introduced Patrick Davis who manages the Energy Efficiency and Renewable Energy Vehicles Technology Program at the Department of Energy. Mr. Davis is responsible for two major government-industry partnerships: the Freedom Car and Fuel Partnership and the 21st Century Truck Partnership. The DoE's Vehicle Technologies Program is the major source of funding for the electric vehicle effort.

He also welcomed Dr. Grace Bochenek, director of the U.S. Army's Tank Automotive Research, Development, and Engineering Center (TARDEC) in Warren, Mich. TARDEC, which he said "is recognized as the ground vehicle center of excellence and the premier laboratory for advanced military automotive technology. Dr. Wessner also welcomed the third panelist, Dr. John Pellegrino, who directs the sensor and electronic device program at the Army Research Laboratory.

THE DEPARTMENT OF ENERGY PERSPECTIVE

Patrick B. Davis
U.S. Department of Energy Vehicle Technologies Program

Although several people have lauded the DOE for moving quickly on the advanced battery funds, "what was important for us was to do this right," said Mr. Davis, who runs the agency's Vehicle Technologies Program. "From the start, we really viewed this as an once-in-a-lifetime opportunity."

Mr. Davis noted that he started working on battery technologies 30 years ago. "During my career, never has an opportunity like this come up, to establish an advanced battery manufacturing capability in the U.S.," he said. "So for us, it was all about getting it right, not blowing this once-in-a-lifetime opportunity."

The DOE began working with the National Academy of Sciences very early in the process, "to start the ball rolling to get to where we are today," Mr.

Davis explained. "It is not just about manufacturing something and throwing money at the problem. We also want to make sure that we're doing everything we can from a technology standpoint." The agency also has been working with modeling "to help make sure we succeed," he said.

The overarching goals of the Vehicle Technologies Program are reducing petroleum dependency and mitigating carbon, Mr. Davis explained. The Recovery Act "layered on top of that the idea that we are trying to stimulate the economy and domestic jobs, and to do it pretty quickly," he said.

Two-thirds of petroleum used in the U.S. is in the transportation sector, Mr. Davis noted. Consumption in that sector has grown from around 7 million barrels of petroleum per day in 1970 to around 14 million in 2010. Use is projected to near 17 million barrels in 2035. He presented a chart showing that petroleum use for air, heavy trucks, light trucks, and cars all are projected to keep rising. However, U.S. domestic production of oil has dropped by more than 40 percent, to less than 8 million barrels per day, since 1970, and is not expected to increase by much.

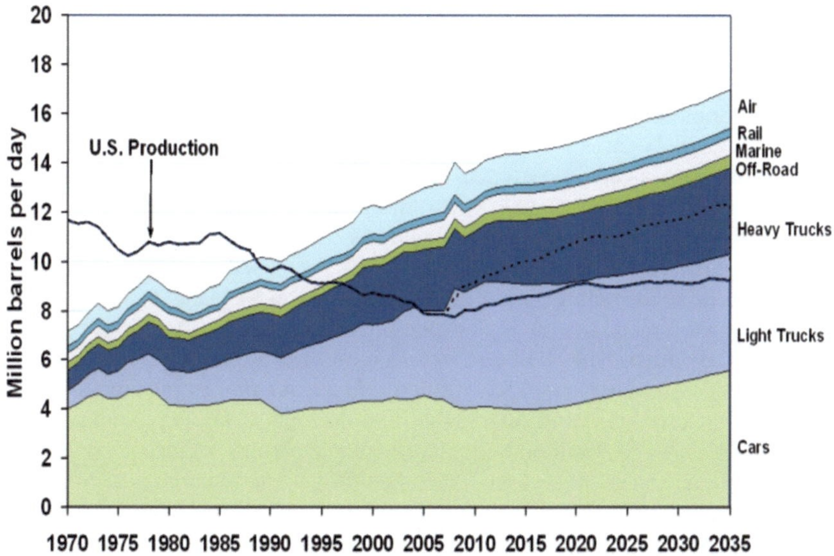

FIGURE 1 U.S. Petroleum Production and Consumption, 1970-2035.
SOURCE: Stacy C. Davis, Susan W. Diegel, and Robert G. Boundy, Transportation Energy Data Book: Edition 30, Oak Ridge, TN: Oak Ridge National Laboratory, June 2011.

Two-thirds of the petroleum consumed in transportation is used in on-road vehicles, Mr. Davis added. There now are 240 million vehicles on U.S. roads. An average of 15.7 million new cars and light trucks were added annually from 2002 through 2007, while 13 million such vehicles were taken out of the market.

This data suggests it will take a long time for electric vehicles to make a national impact. "It takes time for a new technology to take over a market," he said. "And after you have achieved maximum market share, it takes a lot longer than that to essentially replace the vehicles that are already on the road. If you are trying to realize maximum benefit from a new technology introduced today, it takes three or four decades to get to that point. Petroleum is a very serious problem, but it takes time to solve."

Transportation accounts for about one-third of greenhouse gas emission, Mr. Davis noted. Over the past year, the DOE has developed projections of emissions for different kinds of vehicles through 2030. The aim is to look at "what types of vehicles produce what kind of benefits in terms of greenhouse gas emissions and petroleum," he said.

The vehicles roughly fall into three categories, he said. The first group of vehicles uses conventional technologies and some hybrids. They emit an average of 430 grams of CO2 equivalent per mile. The middle group of vehicles includes plug-in hybrids or that are advanced but run on conventional fuels. They emit less than half the CO2, between 150 and 200 grams per mile. The last group consists of electric-drive vehicles or ones using renewable fuels such as hydrogen. Emissions in that group range from around 50 grams to 130 grams. "Those are the only cases that really achieve drastic reductions," Mr. Davis said. "So electric drive is very important to solving those overarching problems."

In terms of budget, the DOE has significant ongoing activities in electric vehicles that go beyond the programs funded by the Recovery Act, Mr. Davis explained. It has a $101 million budget in Fiscal Year 2010 for R&D in batteries and electric-drive technology—with $75 million of that focusing on batteries. The request for FY 2011 is for $121 million. Of that, three-quarters is to be dedicated to batteries. The DOE also has a $44 million program for vehicle and system simulations and testing, $57 million for research on combustion, $51 million for materials research, and $24 million for fuels technology. In addition, the DOE also has a $33 million budget for deploying technologies.

In all, half of the DoE's annual budget for vehicle research is devoted to electric-drive technologies "of one kind or another," Mr. Davis said. What's more, the agency's Vehicle Technologies Program has seen its annual budget increase by 50 percent since 2008.

These are only some of DOE programs, however. The Recovery Act allocated $2.8 billion to advancing electric vehicles, Mr. Davis noted, with about $1.5 billion of that dedicated to batteries. The Advanced Technology Vehicle Manufacturing Loan Program has included awards that involve electric drive, he said. Other critical work at the DOE includes programs by the Office of Science, ARPA-E, and the Office of Electricity.

The DOE has set very aggressive targets for batteries. "We are a very target-driven organization," Mr. Davis said. Much of these targets regard cost. As a benchmark, hybrids such as the Toyota Prius in 2008 cost $1,000 to $1,200 per kilowatt-hour. Vehicles like the Chevy Volt in 2010 are expected to cost $700 to $950 per kilowatt-hour. Currently, the DOE estimates the average cost at about $800 per kilowatt-hour, but Mr. Davis added that if "you ask 10 different people you get 10 different answers."

The goal is to get the cost down to $500 per kilowatt-hour in 2012 and $300 in 2014 for a plug-in hybrid. He pointed out that these standards are meant for laboratory demonstrations. "We're not saying you are going to be able to go out and purchase one at that cost at that time," Mr. Davis said. But tests will demonstrate that the technology could cost that much if vehicles are produced in high volume.

Estimates of the future electric-vehicle batteries market vary widely. Some $8 billion in lithium barriers were made in 2009. "But if you think about it from an order of magnitude basis, the lithium battery market today is pretty big but is all based in consumer electronics," he said. "And it is largely based in Asia," he said.

Today's market for hybrid vehicle batteries is dominated by nickel metal hydride, Mr. Davis noted. Assuming that some 50,000 hybrid electric vehicles are sold a year and that the average battery pack costs $3,000, that makes the market worth only $1.5 billion.

Projecting into the future gets "really sketchy," he said. One study estimated the market in 2015 could be worth $8 billion, assuming that 800,000 electric vehicles will be sold and that the average battery packs will cost $10,000 each.[11] Another study, which projected 6 million electric vehicles would be sold in 2020 at an average cost of $5,000, fixed the market at that time at $30 billion.[12] "I'm certainly not standing up here saying this will be so," Mr. Davis said. "But even if they are off by a factor of five, you still are talking about a large market. You are talking about something that is significant not only from an economic standpoint but from a jobs standpoint."

The U.S. has its work cut out to achieve global market share. The U.S. currently produces only about 1 percent of lithium-ion batteries, Mr. Davis pointed out. China accounts for 25 percent, South Korea for 27 percent, and Japan for 46 percent.

The DoE's strategy through the Recovery Act was to establish the complete value chain for advanced batteries, Mr. Davis said. "We knew from the start we didn't just want to look at cell production or battery assembly," he said. "It was about the entire chain, everything from electrodes to separators to supplying a cell-manufacturer who then supplies a battery assembler who then delivers to an OEM. That was important from the start."

---

[11] See H. Takeshita, 26th Battery Seminar, Ft. Lauderdale, Florida, March 2009.

[12] Estimates from Roland Berger Strategy Consultants and Pike Research 2010.

The agency is not trying to convey that all of this funding "in and of itself establishes an industry," Mr. Davis said. "It doesn't. It is seed money, really." Some people may regard $1.5 billion as a lot of money, he said, especially since that is matched by private investment that brings the total to $3 billion. But that capacity will only supply about 5 percent of the existing vehicle market, "and in the long term we want to do much better than 5 percent," he said. "So we have looked at this from the start as the money that will help get an industry started. We certainly hope that industry will grow, and we honestly expect it to grow without further government funds to support the build-out of it."

The DOE began soliciting funding proposals under the Recovery Act on March 19, 2009, and announced 48 winners on August 5, 2009. "We are really pleased that every one of those projects is signed, and work is underway today," Mr. Davis said.

The Recovery Act funds were spread across the entire battery manufacturing supply chain, Mr. Davis said. While most funds went to makers of cells and battery packs, which require large production facilities, they also support "key supplier industries," he said. Chemetall Foote Corp., for example, received grants to produce lithium. Three companies received funds to make cathode material and three others to make anode materials. A grant also went to a recycler of lithium ion.

In terms of manufacturers, most of the funds went to factories for lithium-ion batteries. They include Johnson Control plants in Holland, Mich., and Lebanon, Ore., involving $600 million in investment; A123 plants in Romulus and Brownstown, Mich.; a $191 million Saft America plant in Jacksonville, Fla.; and EnerDel's $180 million plant in Indianapolis. Many of these facilities use different battery chemistry technologies.

The DOE funds are not limited to lithium-ion. Its funding announcement was for "advanced batteries," Mr. Davis noted. The DOE is interested "not only in whether the technology could do what it needed to do in a vehicle, but also the ability of the companies to succeed," he said. "Did they have customers lined up? What was their viability?"

Non-lithium projects receiving funds include East Penn Manufacturing Co.'s $98 million plant in Lyon Station, Penn., and Excite, which is making cells and batteries in Bristol, Tenn., and Columbus, Ga. Both companies are doing work in advanced lead acid technology. "The reason they were awarded is that we thought there was a compelling case that this technology could be used in the micro-hybrid market," Mr. Davis said.

Almost all of the battery manufacturing plants are located east of the Mississippi River. That is because governments in those states offered incentives that convinced companies to locate there, he said. When all of the facilities are in operation, they will be capable of making around 500,000 batteries a year. The estimate assumes that the average battery will produce the equivalent of 10 kilowatt hours of power.

The DoE's Advanced Vehicle Technology Manufacturing Program offers grants to projects that will add capacity for another 480,000 batteries. Projects funded by both programs will boost production of advanced batteries from 50,000 units to nearly 300,000 in 2012. By 2015, "you are talking about a capacity of almost 1 million batteries per year," he said. The AVTM program facilities, however, focus on batteries for electric vehicles that would be larger than 10 kilowatts each, he added.

Another DOE program that will advance the industry funds transportation electrification demonstration projects. So far, eight grants have been awarded under the program, which Mr. Davis said is the "largest-ever coordinated electric-drive vehicle and charging infrastructure demonstration effort" in the U.S., he said. These demonstration projects and others will lead to deployment of 10,000 electric-drive vehicles and chargers. The vehicles include medium- and light-duty trucks and heavy-duty passenger and commercial vehicles that will operate in a variety of climates and environments. The chargers included devices for the home and some public charging units.

The transportation electrification program studies consumers. "We are very interested in how people will use these vehicles," Mr. Davis said. It is assumed that people will charge vehicles at home every night. "We're not exactly sure people will do that," he said. "How often will they actually charge a vehicle? Where will they charge it? Those kinds of questions are important for commercialization of this technology."

Education and outreach are other DOE priorities. The goal is to encourage and support local and state governments to adopt practices that advance the energy, economic, and environmental security of the U.S., Mr. Davis explained. The DOE has awarded 10 grants to education programs for everyone from grade-schoolers to university graduate students and first-responders and technicians, he said.

The DoE's Clean Cities program works with 86 active coalitions in 45 states. It has been working with alternative vehicles for 15 years. More than 2,000 hybrids and electric vehicles and 1,600 charging stations have been deployed through the Clean Cities program, Mr. Davis said.

The government's Smart Grid initiative also is important to the rollout of electric vehicles, Mr. Davis said. The DoE's Office of Electricity manages a program that has invested more than $8 billion, both in federal and non-federal funds, in more than 100 projects. These programs so far have led to deployment of more than 18 million smart meters that now are being used by 13 percent of America's 142 million electricity customers. They also are responsible for 100 plug-in hybrid charging stations, 176,000 load control devices, 206,000 "smart transformers" that allow for preventive maintenance, and 671 automated substations that account for 5 percent of the 12,466 transmission and distribution substations in the U.S.

Smart grid will not be very important for the rollout of the first electric vehicles in the fall of 2010, Mr. Davis said. "But when you start talking about a million vehicles, and we certainly hope to do that in this decade, smart grid becomes very important pretty quick."

THE ARMY PERSPECTIVES

Grace Bochenek
U.S. Army Tank Automotive Research, Development and Engineering Center

Advanced batteries are very important technology for the military's future and for the future of our nation, said Dr. Grace M. Bochenek, Director, U.S. Army Tank Automotive Research, Development and Engineering Center, better known as TARDEC.

"TARDEC's prime mission is to get the best capability we can to those soldiers who go into harm's way every day on behalf of all of us. We work very diligently to make sure we are building the next-generation systems and the best capabilities we possibly can." Dr. Bochenek explained.

Within the Army command structure, TARDEC's R&D programs work under the Research Development Engineering Command, known as RDECOM. At the U.S. Army Research Laboratory, Dr. John Pellegrino leads a technology focus team. Dr. Pellegrino brings together all of our capabilities to refocus power and energy. The R&D operations bring specialists from different disciplines together to solve complex problems, while Dr. Pellegrino looks at power and energy challenges across the full spectrum. My role is to determine how to integrate that technology onto platforms and to help shape investment strategies, Dr. Bochenek explained.

TARDEC has a full spectrum of R&D and systems engineering responsibilities for the entire life cycle of Army ground vehicles. With more than 500,000 vehicles, the U.S. Army maintains one of world's largest fleets, she noted. One of TARDEC's most important tasks is "developing next-generation capabilities" for this vehicle fleet, Dr. Bochenek said.

Transforming energy use is a "large strategic goal" for the Army, Dr. Bochenek said. Its priorities are explained in the Army Energy Security Implementation Strategy[13] published in 2009. The document discusses cutting fuel consumption, boosting energy efficiency "at the platform and installation level," increasing use of renewable energy, increasing access to petroleum and other resources, and reducing adverse effects to the environment.

---

[13] See Army Senior Energy Council and the Office of the Deputy Assistant Secretary of the Army for Energy and Partnerships, "Army Energy Security Implementation Strategy," Department of the Army, Jan. 13, 2009 (http://www.asaie.army.mil/Public/Partnerships/doc/AESIS_13JAN09_Approved percent204-03-09.pdf)

By improving fuel efficiency, the U.S. Army can use fewer convoys to move petroleum fuel to its ground vehicle fleets, Dr. Bochenek explained. According to one estimate, a $10 increase in the barrel of oil can translate into $1.3 billion in added costs to the Defense Department. While the economic drivers are substantial, the force structure and force protection impacts are even more important to the Army, she said. The biggest end items the Army moves on the battlefield are fuel and water. "When we use logistics convoys to move both fuel and water, it is important for us to attack energy efficiency on all our platforms." In Kuwait, the Army moves around 431 million gallons a year. That translates into 140,000 trucks and 9,300 convoys, with an average of 120 soldiers per convoy. It adds up to 644,000 soldier trips per year, she said. "So if you have fuel savings of only 1 percent, that reduces the number of soldiers you have to put in harm's way by 6,444 fewer soldier trips, which is significant," she said. "Putting it into the perspective of a warfighter adds a different dimension from what we often think about."

One reason the Army's fuel use keeps rising is the kind of equipment it uses on its vehicles, Dr. Bochenek explained. "Those vehicle platforms that used to just carry soldiers in convoy formation from Point A to Point B are now becoming much more complex machines," she said. There are jammers, satellite remote sensing equipment, systems for defeating improvised explosive devices, and active protection systems. "Each of those systems, added with new weapons, adds a new layer of energy requirements at the platform level," she said.

Fuel costs and operational issues also are important considerations. In World War II, the Army consumed about one gallon of gas a day per soldier. Today, it consumes around 20 gallons. Half of that is used to generate electrical power, she said.

Strategically, the Army looks at energy as "a system-of-systems." Dr. Bochenek said. The needs of soldier power, air power, vehicle power, and field/base power are interconnected. "It really is the integration of all of those different assets that we think about in this whole business of power and energy," she said. The Army assesses the full spectrum: "from the way you generate power to how you store it, integrate it, and move it around. We look at the potential for moving around from soldier, to vehicle, to air, to installations."

The major "outcomes" the energy strategy strives to achieve are:

- Enhance ground force effectiveness, flexibility, protection, and freedom of movement by reducing the need to transport fuel.
- Dramatically reduce the sustainment footprint and lighten soldier load and extend platform range and self-power endurance by combining component functions.
- Increase flexibility by expanding capabilities to use alternative energy sources, recycle energy, water and waste, and redistributing resources among systems.
- Reduce the size and number of soldiers and systems required in forward areas by deploying unmanned systems.

- Integrate power and energy situational awareness and management functions with Mission Command to optimize energy use and enable "energy-informed operations."

In the future, the Army aims to reduce fuel-consumption by 40 percent, both in terms of weight and volume. That would increase tactical range and reduce supply demands that are part of the burden of going to war, Dr. Bochenek explained.

One example is the Joint Light Tactical Vehicle program, or JLTV. The Army would like to improve fuel-efficiency with the JLTV, which aims to replace the HMMWV, to 10 Payload Ton Miles per gallon over the JLTV OPS/MP.[14] That represents a 15 to 20 percent improvement in fuel efficiency over the current HMMWV, she said. The Army also wants a 20 percent increase in continuous power available on vehicles. Pulse power-based systems for survivability and lethality power needs range from kilo joules to tens of mega joules, Dr. Bochenek said.

Another goal is to increase fuel economy by 40 percent. The Army wants Abrams tanks, for example, to operate one or two days in combat without refueling, she said. The Stryker armored combat vehicle requires a cruising range of around 330 to 380 miles for a fully loaded vehicle.

Power requirements for vehicles also are rising. New Army vehicles will need anywhere from 10 to 30 kilowatts of on-board power, Dr. Bochenek said. The Army wants vehicles that have an export power capability that would boost their power needs by another 10 to 30 kilowatts. The Army also wants systems that can supply "boost power," enabling vehicles to accelerate quickly. Other goals are for batteries that will enable a silent mobility range of one-quarter mile to one mile for ground combat vehicles and a range of two to eight hours for Silent Watch.

To address such needs, TARDEC has been "investing in understanding this entire range of requirements and how they then translate into vehicle needs, as well as the subsystems necessary to achieve those requirements," she said. It has been developing components such as batteries, intelligent power management systems, solid-state silicon carbine power electronics, high energy-density capacitors for weapons systems, and new fuel cells.

The Army's needs touch on the entire continuum of battery capabilities, no matter if we go to full electrification or to more conventional solutions. That is important as one looks at the entire business case for developing batteries with higher energy and power density along with lower costs, Dr. Bochenek said.

TARDEC uses a "systems approach" to addressing its energy needs, Dr. Bochenek explained. The first level is components, such as motors, batteries, engines, and high-temperature electronics. At the "system integration" level, it

[14] Ton-miles per gallon is a measure of fuel efficiency used in transportation. The measure refers to the amount of fuel needed to move one ton by one mile. Sixty ton miles, therefore, means it would take one gallon of fuel to move 60 tons of freight.

develops concept vehicles and conducts analysis, simulation, and testing. At the "platform level," TARDEC develops demonstration vehicles using new technologies. "My organization over the last 15 years has done a whole host of demonstrators," Dr. Bochenek said. She estimated the Army has invested along with industry in nearly 100 different demonstrators "in order to understand the whole, broad spectrum of electrification." Through this work, the Army has learned that hybrid-electric technology is feasible for meeting many of its needs. "Fuel economy is directly related to the engine operating in the most efficient areas of the fuel map and to regenerative brake energy recovery" she said.

TARDEC has had an ongoing robust testing and evaluation program to understand hybrid systems, Dr. Bochenek noted. These tests show fuel-economy improvements of 10 percent to 15 percent. There still are many issues, however, such as reliability of systems and design optimizations, she said.

To convey the rapid advances in energy storage technology to date, Dr. Bochenek displayed a chart on energy-density improvements over time. In the1860s, lead acid batteries stored around 30 watts of energy per kilogram. Densities rose to 60 watts per kilogram with nickel-cadmium batteries in the 1980s and to 120 watts with nickel metal hydride batteries in the 1990s. Today's lithium-ion batteries produce 145 watts per kilogram, while lithium-ion polymer batteries are projected to offer 200 watts by 2012. Big leaps are expected in 2035 and beyond with future technologies. These higher energy densities will be needed for future vehicles. For example, today's JLTVs require around 15 kilowatts of power, and in the future, they will need around 40 kilowatts. Electrical power needs of the Stryker also will rise dramatically. Future ground combat systems will need nearly 50 kilowatts. "Our Silent Watch and silent mobility requirement really is driving the need for higher energy-density batteries," Dr. Bochenek said. "We really need to increase the power and energy density of the batteries. That is one of the biggest issues we see in this whole area of energy storage and battery technology."

The military's needs present several special challenges for battery technologies, Dr. Bochenek said. One is that "military duty cycles are extremely different from the commercial market, which makes it a little bit of a challenge for us to use solutions that are similar to those of commercial industry," she said. Army vehicles sometimes operate off road, in desert conditions, and on pavement. "Sometimes these hybrid systems are tweaked and refined based on that duty cycle," she said. This also makes it challenging to achieve the desired fuel economies. The military also has "low tolerances for system failure," Dr. Bochenek said. "Reliability and safety are important to us." These needs are especially important for vehicles that rely more on electricity. TARDEC is addressing reliability and the risk of thermal events with new battery chemistries. "We have been working to reduce various types of hazards that might occur," she said. For example, ballistic testing on the cell, module, and pack level has been conducted for lithium-ion batteries. Integration and packaging also are very challenging. "You can't achieve the goals we are hearing about today without understanding how you will integrate them on the platform and the tradeoffs between the various technologies and other requirements," Dr. Bochenek said. Cost is another critical consideration. With the Joint Light Tactical Vehicle program, for example, advanced lead-acid batteries producing about 2 kilowatt hours of power cost $800 to $900 and weigh 180 pounds. The same packaging for lithium-ion batteries produces 3 kilowatt hours and weighs around 60 pounds. They cost around $3,000 to $4,000 each. "So there is a trade there," Dr. Bochenek said. "We need to work really hard to decrease the unit cost over time and at the same time get the payoffs you can get from these advanced battery systems, such as the performance and volume, which are really critical."

In conclusion, Dr. Bochenek said that she thinks electrification and hybridization is "well on its way within the Department of the Army."

John Pellegrino
U.S. Army Research Laboratory

As one looks at the major strategic opportunities for reducing energy use for the military, "you will note that batteries run rampant throughout them in almost every capacity," Dr. Pelligrino of the Army Research Laboratory said.

Dr. Pellegrino listed the major opportunities as follows:

- Tactical unit energy independence
- Autonomous platform power
- Adaptive Power Networks
- Energy Optimized Platforms
- Electric Weapons and High-Power Sensors

The battery technologies must be put together in different ways and each have challenges, Dr. Pellegino pointed out. "But nonetheless, the same kind of technologies work across them," he said.

The military will deploy these technologies in three key domains: The soldier, mobile devices and vehicles, and platforms and weapon systems. A soldier uses tens of watts of power, but weapons systems require tens of thousands of watts, Dr. Pellegrino said. "They are the same technology bases put together in very different ways to enable them," he said.

Tactical unit independence means enabling soldiers and marines to work longer periods of time and being able to reduce convoys, Dr. Pellegrino explained. "If you can be independent for a few days or a week longer, that would be a great boon. New technologies for energy storage and generation are really key to making that happen."

The Army allocates power and energy-technology R&D resources to address four general challenges. Electric power generation and conversion research gets 25.6 percent of the Fiscal Year 2010 budget. Another 27.2 percent is devoted to energy storage, 16.6 percent to power control and distribution, and

30.6 percent to thermal management. These percentages vary in U.S. military services based on their different needs, he said.

The Army has its own technology road map for each battery technology, Dr. Pellegrino explained. "Part of the reason is that we see more extreme environments than the average citizen," he said. "We not only are in more extreme temperature environments, but we also get shot at and have fires. Safety is really important to us. So we need different battery chemistries and technologies."

As the Army invests in new technologies, it can expand upon work done in the commercial sector and offer commercial opportunities, Dr. Pellegrino said. As a result, the battery area is "very, very ripe for partnerships," he said. "The paradigm we have been following in the Army has changed over the past five or 10 years in that we are doing much, much more early collaboration with industry." The Army still awards traditional contracts to corporations to develop technology, Dr. Pellegrino said. But the Army is making greater use of cooperative R&D agreements, partnership intermediary agreements, and "all of the other variety of tools to get government and the universities to engage very early on," he said. "That tends to make the transition go much faster. So that paradigm shift is very, very important."

The Army Research Lab is looking at a number of new technologies. One research project focuses on new electrolytes for higher-density lithium-ion batteries for Army applications, he said. There can be significant commercial benefits as well, he said.

Another research project looks at biologically inspired construction for high-performance anodes for high-power, lightweight lithium ion batteries for light electric vehicles and unmanned aerial vehicles. The bio-inspired construction catalytic synthesis process grows tin nano-particles inside graphite. This can increase the surface area and boost energy density, Mr. Pellegrino explained. It also can prevent disintegration of metal and loss of capacity that occurs with other anodes.

Army labs have collaborated with the University of California at Santa Barbara for years. "It illustrates the new approaches of being very, very multidisciplinary," he said. In order to bring up the next generation of researchers, there is more cross-over between physics, electrical engineering, electrochemistry, chemistry, and biology than ever before, he said. "So training these students, putting them in cross-disciplinary teams, and having them work with the system integrators early on so they understand the issues and technology challenges is important." Another collaboration is with the University of Michigan, Dr. Pellegrino said. Researchers are developing autonomous micro-systems for both mobility and electrical generation.

Battery technology "is kind of at the center of the universe as we go more toward a systems and system of systems approach," Dr. Pellegrino said. There are still challenges, such as the life, endurance, voltage, capacity, and operating characteristics of batteries. "But they play into a formulation where you can start thinking of smart grid applications and sharing of technology among different parts of a base," he said. A simple example is that there are generators at different parts of military bases. "We have neither the devices nor the technology to share energy across the base," he said. "One generator may be working at over-capacity and the other at under-capacity, so we use tons more fuel than we have to."

Bringing smart-grid networking capability together with renewable resources can have a major impact, Dr. Pellegrino said. "While each of these technologies offers a small contribution, together they offer a huge contribution in getting more toward energy independence, either on a small scale or a larger scale," he said.

Partnerships between Army labs, universities, and industry are critical, Dr. Pelligrino said. New Army partnerships cross many barriers and offer" paradigms that just were not there several years ago," he said. These partnerships "are key to bringing it home to those who protect us."

DISCUSSION

To put the discussion into context, Robert Bachrach of Applied Materials noted that the U.S. market for light vehicles is 20 percent of the global market. "We have the largest installed base, but that is not the market," he said. "It may be a used parts market, but it is not where the industry is going." The U.S. military purchases perhaps 10,000 vehicles per year, Mr. Bachrach said. "But we are looking at having to manufacture millions of vehicles a year and millions of battery packs." He said one must "really look at where the U.S. is in the world today, and most of the market is global."

Dr. Wessner asked whether he thought the U.S. can to export to those markets. "Well, I think we have to, don't we?" Mr. Bachrach responded.

Dr. Wessner pointed out that "most countries are willing to export to us, but the other way is harder."

"The world is changing, and we have to get back to exporting," Mr. Bachrach said.

The U.S. Army and the DOD in general "are indeed among of the smallest users of battery technology and electronics technology in the world," Dr. Pellegrino responded. "But we don't want each of those vehicles to cost $1 billion. It is only by leveraging and working with the commercial market in those higher volumes that you spoke of that we are going to be able to do that."

In order to export, however, "the stars have to align a lot differently than they have in the past," Dr. Pellegrino said. "It not only has to be a partnership with the military, industry, and universities across the board. It has to be a partnership as well with policies, taxes, and the whole manufacturing infrastructure together. If you get one piece without the other, it's not going to work."

Dr. Wessner concluded the session by commenting that "not everyone in the world is looking forward to us exporting batteries to them. We have to think hard about that component in the strategy if it is a determining one."

Panel II

The State of Battery R&D and Manufacturing in the United States

Moderator:
Ralph C. Brodd
Kentucky-Argonne National Battery Manufacturing R&D Center

THE BATTERY INDUSTRY PERSPECTIVE

Jason M. Forcier
A123 Systems

The effort to establish a U.S. advanced battery industry "is a global fight," said Mr. Forcier, vice-president of automotive solutions for A123 Systems. "This is not centric to the United States. This is an issue that has to be looked at globally."

By way of introduction, A123 Systems was founded in 2001 in Boston, Mass., and has nearly 2,000 employees globally, said Mr. Forcier, A123's senior vice-president for automotive. The company has been building lithium-ion batteries since 2003. Its initial customer was Black & Decker's DeWalt brand of power tools, "which put us on the map," he explained.

A123 listed in 2009 on Nasdaq with the biggest initial public offering that year. "We have had a lot of press in the past 18 months," he noted. The company now has around 1 million square feet of manufacturing space in China, South Korea, and the U.S. Between 2009 and 2012, it will have invested some $1 billion in capacity, he said.

The core battery technology used by A123 originated in the research labs of the Massachusetts Institute of Technology, Mr. Forcier explained. The company used iron phosphate, which Mr. Forcier said was known to be one of the safest chemistries for lithium-ion batteries. The material enabled A123 to increase energy density to produce large batteries, such as those used in transportation and electrical grid storage.

The company focuses on three broad markets—transportation, power grid storage, and consumer industries. A123 supplies the biggest lithium-powered vehicle fleet, the BAE Systems hybrid bus that is marketed by Daimler. These buses have accumulated over 50 million miles, and there are 2000 of the vehicles now on the road, Mr. Forcier said. A123 also has the world's biggest installed base of electrical grid storage systems in the world, he said. In terms of consumer products, A123 supplies batteries for products "that are applicable to our chemistry," such as those needing long-life power and safety, he said.

A123 has one of the broadest customer pipelines in the industry, Mr. Forcier said. It sells battery cells, modules, and packs to more than 40 programs under development and 20 major customers. In addition to BAE, Daimler, and Black & Decker, customers include Procter & Gamble, Magna, General Motors, General Electric, and Delphi, he said.

The company was "very fortunate" to raise the funds needed to invest in the industry, Mr. Forcier said. "As you hear about all these great new technologies, you really can't go anywhere without about $1 billion in cash available," he said. "That really is the state of the battery industry. If you've ever been in a lithium-ion battery plant, you know that it is a very capital-intensive business."

A123's cash raised most of its cash, $400 million, through its 2009 IPO, Mr. Forcier explained. A123 also received a $250 million DOE grant and a $110 million grant from Michigan. The company is in final due-diligence negotiations with the DOE for an additional $233 million loan. "So we are well-capitalized, and that is important," Mr. Forcier said. It takes up to $200 million to $300 million to build one lithium-ion plant to supply batteries for 20,000 to 30,000 plug-in or electric vehicles.

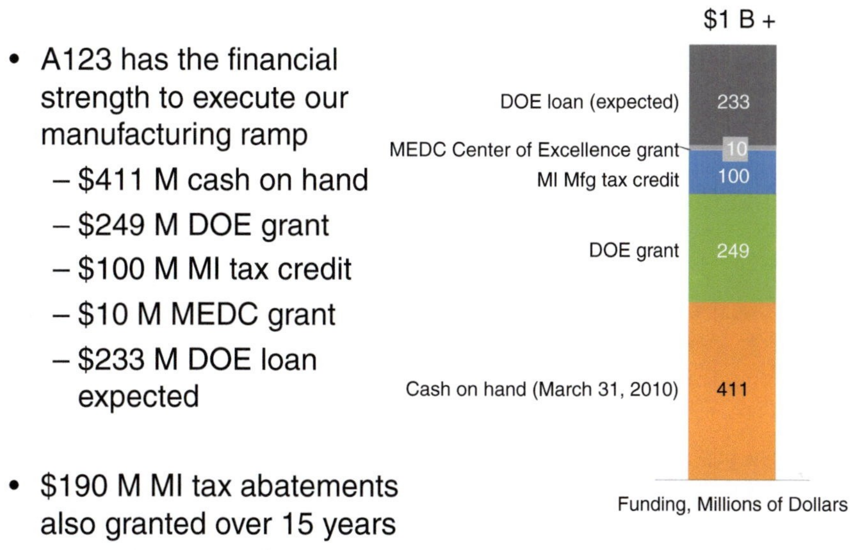

FIGURE 2 Funding for operations.
SOURCE: Jason Forcier, Presentation at July 26-27, 2010 National Academies Symposium on "Building the U.S. Battery Industry for Electric Drive Vehicles: Progress, Challenges, and Opportunities."

The state of Michigan also offered generous tax abatements, Mr. Forcier noted, "although as I told several gubernatorial candidates, we don't intend to pay taxes for another 10 years. If you looked at our quarterly results, you would see that."

In Michigan, A123 has the largest lithium-ion plant in North America, Mr. Forcier said. It invested $230 million in the 300,000-square-foot facility in Livonia. The plant, which produced its first prismatic cells in June 2010, is capable of producing batteries for 30,000 plug-in vehicles or 1 million prismatic cells per month, he said.

A123 also is building a coating plant in Romulus, Mich. Eventually, that will be the site of a "mega campus," where A123 intends to do "everything from powder to coatings to cell manufacturing to packs," Mr. Forcier said.

The big strategic question now facing the battery industry is whether consumers will buy them. "A lot has been done on the creation side. The capacity is in place, and over the next two years a lot of capacity is coming on line," Mr. Forcier said. "So really the key question is about demand."

The price of batteries is expected to come down by 50 percent over the next five years, Mr. Forcier said. Half of that price drop will come as a result of higher production volume. Dozens of new electrification programs are underway across the world just in transportation, he explained. In 2012, "you will see a huge increase in the number of vehicles you buy having electric power trains."

FIGURE 3 Michigan expansion.
SOURCE: Jason Forcier, Presentation at July 26-27, 2010 National Academies Symposium on "Building the U.S. Battery Industry for Electric Drive Vehicles: Progress, Challenges, and Opportunities."

The remaining cost cuts will come through technical advances, Mr. Forcier estimated. “All of us in the industry are working quite heavily on the next generation of chemistry, proving the technology we have, and getting more efficient with our packaging,” he said.

The battery industry is in the “most critical stage in its development,” Mr. Forcier said. Manufacturing plans are getting locked in, plants are being built, and original-equipment manufacturers are deciding to go down certain paths with their technology. “Those are long-term commitments,” he said. “So it is very, very important we achieve most of these improvements now, up front. Doing it five years from now is great, but the industry will really be locked in from a manufacturing perspective. And OEMs will be down the path in committing to certain technologies and architectures.”

In terms of where the battery industry will be based, the competition no longer is only between states such as Michigan, Mississippi, and Alabama, Mr. Forcier said. “This is a case of the United States competing against countries,” he said. “China has a very aggressive subsidy policy. They continue to amaze me with new announcements.” China pays a direct subsidy of $8,800 per vehicle to electric vehicle manufacturers in five cities. Municipal governments have announced credits of up to an additional $5,000 per car, he explained. Shanghai waives license plate fees for electrified cars. The central government also subsidizes makers of electrified vehicles, many of which are partly owned by government entities. To get subsidies, automakers must have a firm grasp on core technologies such as batteries, the electric power train, or electronics.

China requires foreign companies to manufacture in the country if they wish to sell to the domestic market, Mr. Forcier noted. “China takes this very, very seriously,” he said. “Exporting batteries is highly unlikely. You have to build them in-country. China is making sure that happens by the way it is structuring incentives.”

Germany is using a different tack to promote the electric vehicle industry. The government announced a goal of having 1 million electric vehicles on the road by 2020, an ambitious target for a nation with around 4 million total vehicles. “German OEMs are working together quite strongly on standardizations around battery cells,” he said. Germany is “forcing and driving the industry” to localize production in Europe, he said. “Here again, if you want to do business in Europe or Germany, you will have to build in Germany,” Mr. Forcier said. “You will not be able to afford the export value-added taxes and duties that will be assigned to your product. So European business will be won and made in Europe. Asian business will be won and made in Asia.”

- China
  - 60,000 Yuan (~$8,800) EV subsidy will be paid directly to vehicle manufacturers in 5 strategic cities
  - Some of the municipalities have already announced additional incentives ranging from 20,000-40,000 Yuan ($2,950-$5,500)
  - Support is provided to Chinese-owned companies
- Germany
  - Target of 1 Million EVs on the road by 2020
  - In a market of 3-4 Million vehicles annually, the goal is very ambitious
  - Strategic intent is to drive battery development and mfg in Germany

FIGURE 4 Global competition for electrification.
SOURCE: Jason Forcier, Presentation at July 26-27, 2010 National Academies Symposium on "Building the U.S. Battery Industry for Electric Drive Vehicles: Progress, Challenges, and Opportunities."

The key to growing battery sales in the United States is to create domestic demand. "We may not be the biggest auto industry in the world anymore, but the demand has to come from here in the U.S. in order to achieve energy independence and create jobs in the United States," Mr. Forcier said. One way to do that is to electrify the big military and government vehicle fleets. "We think that is a huge opportunity to help stimulate demand in the U.S.," he said.

The U.S. industry does not need stimulus money, rebates and incentives forever, Mr. Forcier said. "What we need are four or five years to get the costs down, to get the models in place, to get battery leasing worked out so that the cost of buying an electric vehicle gets down to the cost of buying an internal combustion engine," he said.

Electric vehicle clearly are cheaper to operate, Mr. Forcier pointed out. "The penalty right now is the up-front cost to acquire the technology," he said. "That is where we need support in the short term." In four of five years, the cost equation will change as battery costs drop by half and "as companies like A123 figure out how to lease batteries and reuse them in secondary applications," Mr. Forcier predicted.

Investments also should be made in infrastructure, he said. "But we have to be focused on infrastructure and demand at the same time." The federal government missed a good opportunity to address such needs through the energy bill, which is struggling to get out of Congress, Mr. Forcier said.

A recent version of the energy bill included a provision for electric cities that would have provided more incentives to buy electric vehicles, he noted. "That is the kind of legislation we need to pass," he said. Mr. Forcier thanked the Michigan coalition in Congress and the state government for supporting such measures. "But on a federal level, it comes down to not having 60 votes," he said.

Mr. Forcier said A123 is happy to be in Michigan and that he believes the industry has a bright future. "But we can't ignore demand," he said. "We've got the creation side covered. Demand is what we really need in order to go forward."

Mohamed Alamgir
Compact Power

The advanced-battery push and symposiums such as this one should have happened 25 years ago, quipped Mr. Alamgir, Compact Power's research director. "Then I wouldn't have had to go through five companies during my career in lithium battery technologies," he said. "If you do a study on what went wrong and right in lithium-ion, you can use me as a case example. This kind of funding was not there before. It was very spotty, which is why we were in trouble."

Compact Power was established in Colorado Springs, Colo., in 2000 to develop large vehicle batteries for LG Chem, Mr. Alamgir explained. When the company shifted to Troy, Mich., in 2005, only three employees came. "If you live in Colorado Springs, the thought of moving to Detroit is not very appealing," he said, noting that his wife has not yet excused him for moving to Detroit. Compact Power's parent company, South Korea's LG Chem Ltd., explained the move by saying (a Korean proverb) , "If you want to catch the tiger, you need to go to the den of the tiger," he said. "That meant the Big Three."

The company now employs 150 in Michigan. Most of the people Compact Power hired knew nothing of batteries at the time, Mr. Alamgir said. "They were very well-educated, so it was not very difficult for them to transition over," he said. "To Michigan's good fortune, many of these guys now are very well informed about battery technologies. And it is very difficult to keep them because there are very lucrative offers to move to other companies."

Funding from the U.S. Advanced Battery Consortium and the Department of Energy "kept this company alive," Mr. Alamgir said. The company went through "very lean times" from 2003 to 2006, he said. "When you went around to companies saying that you have an electric-vehicle and plug-in hybrid battery, they said: 'Come back later. We have no time to address your market now.'" He said he is grateful to the DOE and USABC "for keeping us afloat through those lean years."

Government support has long been important to developers of lithium-ion technology, Mr. Alamgir noted. From 1985 through 1995, he had worked at EIC Laboratories in Boston. That company "survived completely" on funding from the Small Business Innovation Research program,[15] he said. The bulk of Compact Power's initial funds came from the DoE. "I have survived on government funding of the battery industry," he said.

The company addresses "all aspects of the battery pack," Mr. Alamgir explained. It develops battery pack concepts and designs and manufactures packs. It also develops battery-pack management systems, power and signal architectures, thermal management, charge-control algorithms, and test and validation services. It does most of its R&D in-house, but also collaborates with universities. "This is very vibrant research, and Michigan is a beneficiary," he said.

Being part of a large corporation helps companies like Compact Power, Mr. Alamgir said.[16] He noted that LG Chem's parent, South Korea's LG Group, is a huge global conglomerate, with $113 billion in revenue, hundreds of companies, and 160,000 employees. "This shows how deep-pocketed a company has to be to survive in this industry," he said. "You can start a company. But I have seen with my own eyes where you can end up if you don't have enough funding."

He pointed out that 70 percent of LG Chem's revenues come from petrochemicals. Lithium-ion batteries account for just 10 percent of revenues, he said, even though LG Chem is the world's third-largest manufacturer and is widely known for that product. Mr. Alamgir noted that LG Chem's CEO has said that when he goes to parties, people come up to him and congratulate him for his success in batteries.

Rechargeable batteries consume 40 percent of LG Chem's R&D spending, however, compared to just 28 percent for petrochemicals. LG Chem is investing $1 billion in the battery industry over five years.

LG Chem makes lithium-ion batteries in all shapes and sizes, Mr. Alamgir explained. Its biggest business is small cylindrical and prismatic batteries for consumer devices such as mobile phones and notebook computers, supplying companies such as Dell, Nokia, Hewlett Packard, Motorola, and LG Electronics. In the automotive industry, "we are proud that we are the only company that has both Ford and GM as our customers," he said.

---

[15] The Small Business Innovation Research (SBIR) program is administered by the Small Business Administration in the U.S. Commerce Department. It provides early-stage financing for small technology companies.

[16] Compact Power has now been split into two subsidiaries of LG Chem. LG Chem Power (Troy, MI) focuses on R&D of module and pack designs, prototype builds, sales and customer support, whereas LG Chem Michigan (Holland, MI) focuses Li ion cell manufacturing.

Vertical integration is another advantage in the battery industry, Mr. Alamgir said. LG develops most of its manufacturing processes in-house. Because Compact Power is part of a large chemical company, it has access to patented processes and chemistries that are battery-related, he said. Due to the research-intensive aspect of the business, "you have to have a lot of in-house material development and research to be viable," he said.

The DOE and LG Chem each are contributing $151 million for Compact Power's new manufacturing facility in Holland, Mich. President Barack Obama attended the July 15, 2010, ground-breaking ceremony, and the plant is scheduled to begin manufacturing cells in 2012. Initially, electrodes will come from South Korea and be assembled in Holland. Plans call for making electrodes in Holland as well the following year. The plant will have capacity to produce 15 million to 20 million cells, enough for 50,000 to 60,000 vehicles. It will employ around 300 people.

There is some concern in the industry that a "battery bubble" is building, Mr. Alamgir said. Some analysts project significant overcapacity in the industry. Also, some car makers are bringing battery-pack manufacturing in-house. "How does this impact the industry?" he asked.

Numerous companies already have failed in the lithium-ion industry, he noted. The casualty list includes battery maker Duracell, "which in the 1980s was the house to go to for research related to lithium batteries," he said. "It disappeared." After a series of takeovers in the 1980s, Duracell's lithium-ion research operation was dismantled, he explained. Energizer also vacated the business. Other failures in lithium-ion include Polystor, Motorola ESG, Moltech, MoliCell, Electro Energy, Imara, and Firefly, he said, adding that he was part of three companies that disappeared.

Many of these battery companies got into the business at the wrong time. "This is where the government could have helped," Mr. Alamgir said. "I'm sure that some of these guys are saying, 'I wish I were here now.'" At the time, early battery companies could not get enough funding to survive against tough competition from Japanese and Korean companies.

One big lesson from this history is that "the government does need to support research in the future, just as the Japanese government did in the 1990s with their New Sunshine program," Mr. Alamgir said. Even though the U.S. started the Advanced Battery Consortium, "somehow the OEMs and manufacturing companies did not pick up the slack," he said. "Sometimes, jolts like those the stimulus funds provided will play a big role. I think companies like PolyStor or Electro Energy would have benefitted big time from such an investment."

For the long term, however, "visionary and gutsy CEOs and CTOs of big corporations" must support the battery industry, he said. "They have to have the vision to be in this business and think of the common good of mankind, society, and countries," Mr. Alamgir said.

At first, most of the materials for advanced batteries will have to be imported from Japan, South Korea, and China, he said. "We do not have them here," Mr. Alamgir said. "Even though we are working closely with new materials here, it will take two years to bring them in house."

Mr. Alamgir recalled the he worked as an engineer at one battery company that had high-flying customers such as Research in Motion. The company was acquired by Tyco Electronics. One day a division head visited the labs and asked the staff what it did. "We explained we did R&D," Mr. Alamgir said. "He said, 'I don't have any interest in dumping money into research. We have to shut this down.'" Mr. Alamgir said he later found out that the manager came from a profitable Tyco division that made electrical poles. "All you had to do is cut down a tree and make a pole," he recalled. "He did not need research to make money"

The message is that "we need leaders who believe in the future of this industry and are committed to providing funds," Mr. Alamgir said. "Battery research to me is a marathon race. We have a lot of sprinters in this country. We need industry leaders in marathon running like the Kenyans, Somalis, and Ethiopians who can run the race to the end."

THE AUTOMOTIVE INDUSTRY PERSPECTIVE

Nancy Gioia
Ford Motor Company

All major automakers are "dealing with the same set of issues" when it comes to electrified vehicles, said Ms. Gioia, Ford's director for global electrification. "I agree 150 percent that we are in a marathon," she said. "It is a marathon not only of R&D. It is a marathon of new knowledge and manufacturing. It is a marathon that will make a difference to the environment, energy security, and employment in every region around the world."

From Ford's perspective, "electrified transportation" means hybrids, plug-in hybrids, and full-battery electric vehicles, Ms. Gioia said. It means, "any place that we use stored electricity to directly replace oil."

Whether the manufacturer is Ford, Nissan, General Motors, or Toyota, "each company is looking at what is needed for sustainability in the future," Ms. Gioia said. By this, they not only mean their products and fuels, she said, but also the sustainability of their businesses.

In the near term, therefore, automakers will make continuous improvements to internal combustion engines and launch hybrids, Ms. Gioia explained. In the mid-term, from 2011 through 2020, "we will see growth in electrification," she said. "But we also are going to see a number of other technologies continue to improve for petrol and diesel solutions." There will be massive reductions in weight.

The importance of improving traditional technologies should not be overlooked, she suggested. "If you think about it for a sustained business, this has the greatest impact for reduced oil consumption for the largest number of customers with the best use of capital, equipment, and depreciation," Ms. Gioia said. These gains can be accomplished faster with existing technologies because they do not require different infrastructure. "It's just a pragmatic reality. It's not that we are vested in what we already have," Ms. Gioia said. "But if you want to shift and reduce fuel consumption, making your current technologies more efficient and your vehicles lighter also enables improvements in electrification, smaller batteries, and additional efficiencies."

The choice will not be either electric or petroleum-based technologies. "It will be a combined effort going forward," Ms. Gioia said. "There will be fuel diversity and growth in bio-fuels. There is no silver bullet. It is not one answer. It will be a set of answers."

Another aspect of a sustainability strategy is that a company must embed something in its fabric, Ms. Gioia said. "Electrification as a change of technology does not happen until a company embeds it in its resources, its R&D, its capital allocation, and its product development processes," she said. "This is what we have now done at Ford with this plan. I think the companies that will continue to drive this on a global basis have it embedded in their decision-making process as one of the core foundations going forward. That also means it must be a money-producer and provide returns for our shareholders."

Ford is launching products across several technologies. It has the Fusion and Escape hybrids. The Ford plug-in project, which is supported by the DoE, several utility companies, and the Electric Power Research Institute,[17] has been on the road since 2007, she said. In 2010, Ford launched the Transit Connect line of fuel-efficient small commercial vehicles, and it will launch Focus Electric in 2011.

Affordability is the "Achilles heel" of hybrids, which are the foundation of Ford's electrification program, Ms. Gioia said. Ford is attacking affordability on two fronts: by working on battery technologies and by electrifying its "highest-value platforms," she said. Some competitors have created unique platforms for hybrids, she explained.

These major platforms include the global C platform for the Ford Focus sedan. The Ford C Max and Ford S Max multi-purpose vehicles sold in Europe use the same platform, Ms. Gioia explained, as will the Transit Connect line of trucks in the future. Another major platform is the Ford CD, which is used for such midsized vehicles as the Fusion, Milan, and Mondeo. Ford can put up to 10 "top hats," or different upper-body designs, on a single platform, she said. "We can put the technology on a variety of top hats very quickly in global, flexible manufacturing facilities," she said.

[17] The Electric Power Research Institute (EPRI) is an independent, non-profit company performing research, development and demonstration in the electricity sector.

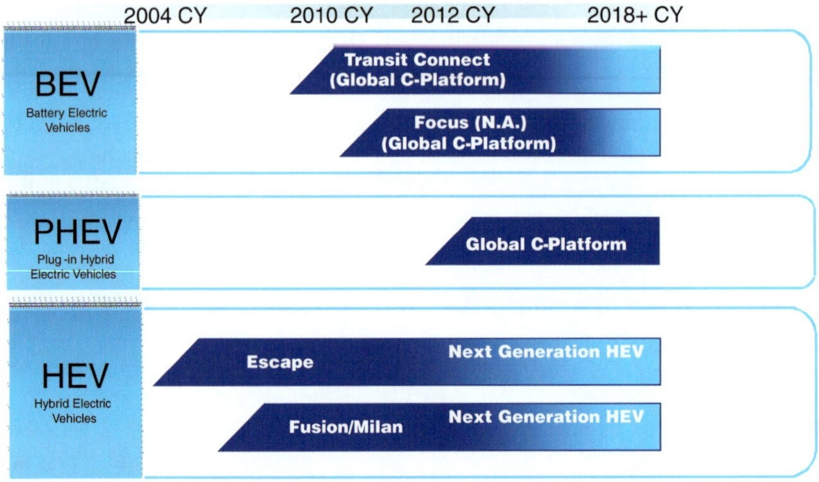

FIGURE 5 North America—announced electrification projects.
SOURCE: Nancy Gioia, Presentation at July 26-27, 2010 National Academies Symposium on "Building the U.S. Battery Industry for Electric Drive Vehicles: Progress, Challenges, and Opportunities."

Electrified transportation, therefore, is not only about the battery. "It is the design, the development, the validation, the prove-out, and the manufacturing processes down the same assembly line," Ms. Gioia said.

Ford is deploying the same electrification strategy globally. It is electrifying its products sold in Europe, Ms. Gioia said, and is looking at doing the same in Asia. "What we want is for our global volume to be electrified quickly," she said. "But we also recognize the reality that transportation must be affordable in each region around the world. So we believe that balanced growth must provide the flexibility to react to volatile market conditions."

Electrified cars accounted for only 1 percent of Ford's sales in 2010, Ms. Gioia said. The goal is for that to reach 2 percent to 5 percent in 2015 and between 15 percent and 25 percent in 2025. Even then, 70 percent of Ford's global fleet will likely be hybrids; up to 25 percent will be plug-ins, and the rest full electrics. The projections acknowledge what it will take to roll out the infrastructure and make it accessible, as well as improve battery technologies to meet customer requirements, Ms. Gioia said.

Ford is hardly alone in believing in a balanced approach. She noted analyses by JP Morgan, Credit Suisse, Boston Consulting Group, A. T. Kearney, and Roland Berger. According to projections based on a compilation of these five studies, hybrids will account for 3.1 million of the 4.3 million electrified vehicles expected to be sold in the U.S. in 2020. Five hundred thousand of those vehicles will be entirely battery-powered.

In Europe, by contrast, less than half of the projected 5.3 million electric vehicles will be hybrids. Plugs-ins will account for 1.6 million and all-battery-powered for 1.4 million of those vehicles. The results will vary tremendously by region based on factors such as fuel prices and government policy, Ms. Gioia said.

Access to intellectual property and the capabilities of the competition also can influence the results. "And there is an enormous amount of competition," Ms. Gioia said. She noted that the projections she cited are only for major automakers that have announced goals of selling 50,000 or more electric vehicles by 2015.

The electric product mix also varies dramatically by automaker, Ms. Gioia pointed said. Toyota, Hyundai, and Honda, for example, expect all of their electric vehicles sales to come from hybrids in 2020. GM projects that more than 60 percent will come from hybrids. Renault/Nissan project that nearly 80 percent will come from all-battery electrics.

Some differences are due to the fact that some companies lacked hybrid technologies, Ms. Gioia explained. Ford and Toyota have locked up much of the intellectual property related to the parallel power-split system that allows cars to run on both the battery and internal combustion engine, she noted. To purchase or get access to that technology, manufacturers must go to Ford or Toyota. Battery electrics, therefore, are emphasized by companies whose electric vehicle-programs are not as established because less intellectual property is tied up, she said. "Another inhibitor is just time, experience, and the complexity of these systems," she added.

To sustain a real mass market, better batteries are of course needed, Ms. Gioia said. Automakers also must still deliver great features. The technology must be "functional and trustworthy," meaning the cars are durable, reliable, and "something I can count on to carry my family or run my business with," she said. An electric car "has to deliver the basic transportation needs. If I have to haul, pull, load eight people into a vehicle, it still has to do that," she said.

The vehicle also must be affordable over time. The cost of electric battery packs should decline from an average of around $750 per kilowatt hour in 2012 to $250 in 2020, Ms. Gioia said. If the range of the battery system is held constant at about 100 miles, the cost gap between battery packs for hybrids and plug-ins are expected to nearly disappear, she said. "That is because the cost of the batteries start to equal the cost of the battery pack that you eliminate by going full-battery electric," she said.

These factors mean consumers will have a lot of choice there will be no clear solution or timeline for rolling out of technologies, Ms. Gioia said. Ford will introduce a full line of electric vehicles. In Europe, it already has introduced

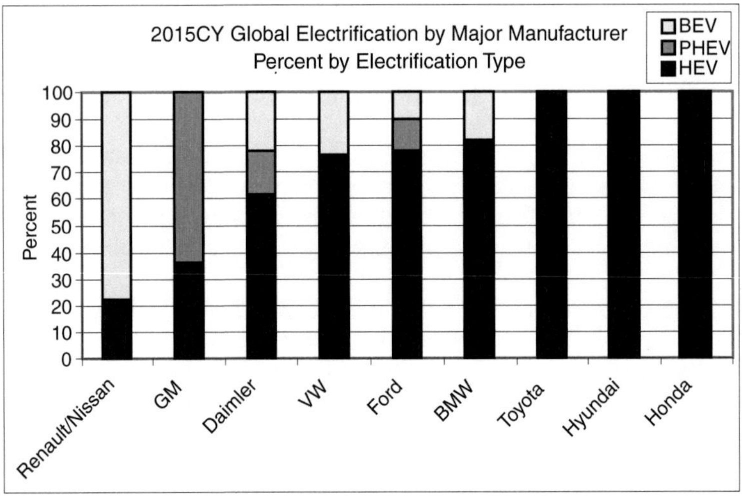

Note:
- All data is from CSM Worldwide global comprehensive vehicle production and sales forecasts, 3/05/10.
- Major manufacturers are those with >50,000 electrified vehicle sales projected in 2015

FIGURE 6 2020MY global electrification volume projections by region.
SOURCE: Nancy Gioia, Presentation at July 26-27, 2010 National Academies Symposium on "Building the U.S. Battery Industry for Electric Drive Vehicles: Progress, Challenges, and Opportunities."

simple start-stop systems, in which a battery starts the engine but does not power the car, that provide 3 percent to 6 percent in fuel savings. Ford also will introduce "mild hybrids" that modestly help drive the car, medium hybrids, and full hybrids, which reduce fuel by 55 percent and have all-electric drives. Ford also will launch plug-in hybrids that save 80 percent of fuel and full battery-powered cars.

Developing the supply industry is critical to the success of electric vehicles. Batteries are not the only challenge. Electrified vehicles need systems to convert alternative current to direct current, regenerative brakes, inverts, and new motors and transmissions, for example. Chargers for electric vehicles are "ridiculously expensive today" and are being developed by "what was a cottage industry," Ms. Gioia said. "As it becomes main stream, we need main stream companies jumping into that with capital and assets to get the cost of those chargers down. So a lot has to happen to make the electrified industry progress."

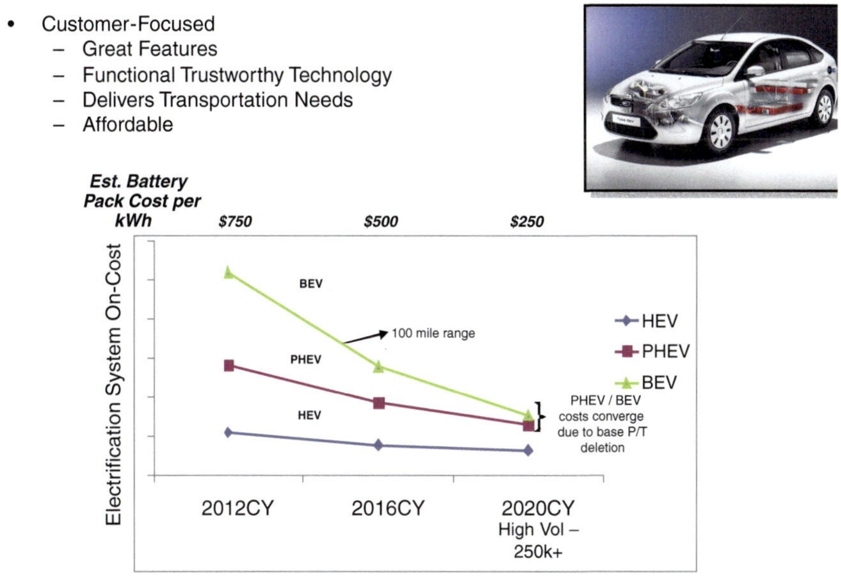

FIGURE 7 What does it take to support a sustainable mass market electric vehicle?
SOURCE: Nancy Gioia, Presentation at July 26-27, 2010 National Academies Symposium on "Building the U.S. Battery Industry for Electric Drive Vehicles: Progress, Challenges, and Opportunities."

An example of new components is the SmartGuage[18] instrument panel on new Fusion hybrids. "This is a whole new world. It's a display, not a video game. You still want people watching the road," Ms Gioia said. "The bottom line is that we need a new set of engineers thinking about how to communicate coaching information with customers real-time to get the most energy efficiency out of their vehicle as possible."

Moving to the "electric state" will involve much more than the transportation and utility sectors, Ms. Gioia said. "We are going from independent systems to integrated systems," she said. "It is a new energy ecosystem. It is a series of industries now coming together, collected in a very complex system. It is important to understand that one element of that system cannot succeed without the other."

Each industry will require deep understanding of other industries it never had to understand before, she said. As an electrical engineer herself in the auto industry, Ms. Gioia said she "never thought about power generation and distribution down to the local transformer to say whether my car will work."

---

[18] SmartGuide is a liquid-crystal display panel used in the 2010 Ford Fusion hybrid and Mercury Milan hybrid that features an "EcoGuide" with animated messages about good driving habits and fuel-saving tips.

Car dealers will have to explain to customers what they must to do to their homes, for example. A plug-in vehicle doubles the energy load of a household while it is charging, she said. "Where do our customers get that information?" Ms. Goia asked. "It is a new knowledge system, and to make it work for the customer it has to work simultaneously as this transportation rolls out."

The charging infrastructure also must be worked out. The top priorities are setting up charging systems at homes, at depots for fleets of vehicles, and at work, Ms. Gioia said. Charging at public spaces is a lower priority. She noted that at a recent session hosted by the DOE to discuss infrastructure needs, "there was fair alignment around the OEMs" about the priorities essential for meeting customer needs.

Charging infrastructure must make it easy for car owners to charge overnight. Different levels of charging systems are needed for different kinds of vehicles. Level 1, in which cars can be charged with a home electrical socket, may work fine for small, low-capacity plug-in hybrids and require investments ranging from nothing to $200, Ms. Gioia said. Level 2 systems, however, will be required for all-battery electric vehicles. These charging systems around $2,000. Level 3 systems for workplaces or public stations can cost $50,000, she said. Even assuming costs drop to $25,000 "that's a heck of a capital investment communities have to make and then maintain," Ms. Gioia said.

For the batteries, different cells are required for different applications. "We don't simply install capacity for one battery and that works for all," Ms. Gioia explained. Hybrids require more power. Plug-in hybrids need both more power and energy density. Full battery electrics require a much higher energy density.

These batteries will evolve. The current battery for the Focus full-battery electric car produces 23 kilowatt hours, adds 500 pounds to the vehicle, and is 125 liters in size. "That is whopping big to fit into a car," Ms. Gioia said. Second-generation batteries for electric cars, which will be available two to three years after the first generation, will weigh around 400 pounds and be 100 liters. But they still will provide a range of 100 miles, she said.

Third-generation batteries, which will come in another two to three years, will weigh 250 pounds and be 75 liters big. "The goal is to make it on par with the fuel tank," she said. "So we need two to three generations of technology before the batteries become truly replaceable in terms of weight, size, and displacement to provide the equivalent 100-mile range. If it turns out customers demand ranges of 200 miles, "that just exacerbates this challenge," she said.

Temperature control, energy density, the number of real-world charge and discharge cycles, and cost also remain significant challenges for full electric cars, she said. "We need to go through two to three cycles of innovation and then scale up appropriately to have a customer-driven product that would be affordable," she said.

The U.S. Advanced Battery Consortium and the battery industry as a whole generally agree much more must be done before electric vehicles are ready for the mass market, Ms. Gioia said. This does not mean Ford won't launch the Focus Electric in 2011, Ms. Gioia said. "Of course we are," she said. "We also are launching Transit Connect later this year."

"Mass market" means moving from 2 percent of car sales to 5 percent, she said. It also means, "we have affordable solutions for other than early adopters or the policy-incentivized world." For hybrid electric cars, batteries are expected to cost $20 to $30 per kilowatt hour per cell in 2012, Ms. Gioia said. Cells for plug-ins will cost $500 to $1,000. The price range varies, due to assumptions in R&D, capital depreciation, labor, and other mark-ups, she said. Lithium-ion cells for laptop batteries are much cheaper because they are produced in mass volumes and because the transportation sector is more demanding, she said. Moreover, all components of lithium-ion batteries for cars—the cathodes, anodes, electrolytes, hardware, and separators—require improvements.

For the U.S. to be fully competitive and not remain simply an importer, "the U.S. battery industry must have world-class and leading technologies," Ms. Gioia said. That means not just in chemistry and materials, but also in manufacturing processes and equipment, she said.

Thanks to government incentives, capacity is now being installed in the U.S. "The knowledge to build the equipment, set the details, and design the processes and equipment for the future is not being brought here yet," Ms. Gioia said. "We need to work on that." Manufacturing processes and equipment will deliver the needed cost reductions. "Without that, the capacity will be underutilized," she said. Japan and South Korea are still the leaders in manufacturing technology, she said.

The U.S. also needs a fully competitive cost structure, Ms. Gioia said. Labor cost is not the big driver. "It is all of those elements and the manufacturing process capability along will scale," she said.

"At the end of the day, there is a lot to be done," Ms. Gioia said. "It requires a tremendous amount of system thinking, with the public and private sectors working together."

THE UNIVERSITY/ STARTUP PERSPECTIVE

Ann Marie Sastry
University of Michigan and Sakti3

After Mr. Alamgir's "chilling tour of the graveyard of battery companies, I will try to be a little more positive and uplifting," said Dr. Sastry, who heads the Advanced Materials Systems Laboratory at the University of Michigan and is CEO of the Ann Arbor-based advanced battery developer Sakti3.

Besides climate change, another driver of electric vehicles is the growing concentration of the world's population into megacities with 10 million inhabitants or more, Dr. Sastry said. Some cost estimates of batteries are pegged to the present state of technology, power, and energy density. "Unless we go very strongly toward something like 500 watt hours per kilogram and 500 watts per kilogram in energy density, it is very unlikely that we will upend some of these limits and see large degrees of electrification," she said.

Economies of scale probably won't be achieved in electric-car batteries until production reaches 300,000 a year, Dr. Sastry predicted. At that point, the cost of a lithium-ion cell for car batteries is projected to drop from more than \$500 now to around \$100. She said it is important to remember that battery properties themselves will enable the car market to arrive.

One problem is that the U.S. lacks the workforce to support such an industry. Dr. Sastry cited a comment by Wanda Reder, president of the Institute of Electrical and Electronics Engineers' Power & Energy Society. "The current graduation rate from U.S. university electric power engineering programs is not sufficient to meet our nation's current and future needs," Ms. Reder said.[19] Studies by other organizations reach similar conclusions, Dr. Sastry said. "We're lacking the people to do this," she said. "The workforce education issues are profound. It's clear that we need more scientists and engineers, and I hope everybody here in your industrial and research efforts also will put efforts back into workforce training. It's not just a good thing to do. It's an absolute requirement for a sustainable business." Dr. Sastry noted that all organizations in the battery industry are facing a challenge now in finding the right workers. "So it is important to join in collaborative activities, because the technology pain is intense right now," she said.

Research in advanced batteries had been underway for more than a decade at her lab at Michigan, Dr. Sastry noted. That research provided "the numerical underpinnings for the work we are doing now in optimizing batteries," she said. That research also trained the scientists and engineers who now are needed in the industry, she said.

A change occurred in 2004. At the time, Dr. Sastry was doing research on nickel-metal hydride and lithium-ion batteries. For the first time, lithium-ion cells became cheaper than nickel-metal hydride. "It had nothing to with magic or any inevitable economy of scale," she said. "It had to do with capability. Lithium ion technology was disruptive." The technology allowed camcorders to work three hours without recharging, rather than a few minutes. "That enabled large markets, which enabled people to do manufacturing research. That improved the cost structure."

Current projections suggest the U.S. market for electric cars will be able to reach the 300,000-unit thresh hold needed to push lithium-ion battery costs below \$300 per kilowatt, Dr. Sastry said. That is the point at which many assume the market will take off. "Manufacturing technologies that do not offer

[19] Amy Fischbach, "Engineering Shortage Puts Green Economy and Smart Grid at Risk," Transmission and Distribution World, April 21, 2009 (http://blog.tdworld.com/briefingroom/2009/04/21/engineer-shortage-puts-green-economy-and-smart-grid-at-risk).

the promise of getting cost out of the product really shouldn't be investigated," she said. "The counterpoint to that is that we have to invest very heavily in new manufacturing technologies."

The Advanced Materials Systems Laboratory works with many partners around the world, Dr. Sastry explained. Partners include the DoE, the National Science Foundation, LG Chem, GM Mainz Kastel, AND Technology, Oak Ridge National Laboratories, and Ford. "We are friends with everybody," she said. "It is really important to do that, because all of the partners have a set of particular skills that are necessary to the problem."

Because her lab is connected to a university, it has the power to convene people and make proposals that bring people together, Dr. Sastry pointed out. If companies in the industry are not one of her lab's partners, they should join or find another group to join, she said. "The adjacent areas are very important in regularizing electric vehicles."

The University of Michigan was one of the first to invest in research and education aimed at improving lithium-ion cells and battery packs, she said. Until recently, however, there hadn't been a strong motive for universities and car companies to work together. "We weren't on the cusp of commercializing the technology," Dr. Sastry explained. As the technology improves and the industry grows, "we see greater impetus for these groups to work together."

Dr. Sastry founded the first Energy Systems Engineering program in the U.S. It began with nine students in 2007 and had more than 200 enrolled as of Sept. 1, 2010. "We were very proud of what we accomplished in three or four years," she said. Dr. Sastry recently handed over leadership of that program to focus on other things, she said.

The University of Michigan has joined with GM and the U.S. Advanced Battery Coalition to address all aspects of the electric power train. It conducts basic research to understand why materials fail, for example, and to develop controls algorithms. The ultimate goal is to get those controls algorithms into vehicles, Dr. Sastry said. "So if you do it right, at the vehicle scale you are using computational training that goes all the way to the atomistic and micro scale in the battery cell," she explained. "That takes a lot of different people."

The "technology story is important," Dr. Sastry says, "because it tells you why all of these groups have to work together." The physics of battery chemistries and electrochemical cycling "are not trivial," she explained. "Even though you can write down the kinetics in a straightforward way, the reality is that it is a combination of mechanics, thermal effects, heat transfer, kinetics, and a whole host of other disciplines that are required to build simulations that allow us to say how long a battery cell will live and how well it will cycle." These simulations also predict a cell's capacity and the effect of temperature.

Part of what makes the undertaking difficult is "the science of how to put all those people together and execute," Dr. Sastry said. Michigan, which has more than 70 people involved in the various institutions and national laboratories, spends a lot of time bringing the right people together, she added.

Time scales for the technology also are important, Dr. Sastry said. "People who do computational experimental work have to worry a lot about time constants and how long it takes to derive the parameters that tell us how a system is going to behave." Such battery factors as cycle life "can really only be understood if you understand the scale at which things are breaking down or occurring inside the battery cell," she explained.

Different expertise is needed to work on all of these problems simultaneously, Dr. Sastry said. "You want people who understand diffusion, who understand kinetics, who understand heat transfer and thermal effects, and people who understand mechanics," she explained. "To put these equations and experiments together is not trivial."

Her program has spent more than a decade and millions of dollars to get this far, Dr. Sastry said. The team now can "predict pretty satisfactorily" factors such as capacity, the effects of thermal cycling, and off-gassing, "but we have a long way to go," she said. These numerical simulations influence the cost and choices of technology.

Research in manufacturing systems "that are fungible across platforms" also need support, Dr. Sastry said. "Unless the government funds approaches that can make many types of chemistries, we will fail to develop the variety of battery cells that meet the variety of needs the Army team so ably talked about." Many interesting partnerships will follow, Dr. Sastry predicted. "Big companies will act like small innovators and vice-versa," she said. "Universities and industry will adopt new roles."

Regarding Sakti3, the company she helped found at the University of Michigan, Dr. Sastry noted that Henry Ford started Ford Motor from the winnings of a race. "The immutable dominance of existing big companies is not inevitable," she said. "All of these big companies started out small. And it is something that America is particularly good at doing. And it's something that America relies upon."

It is very important that the U.S. government support innovations coming out of America's national labs and universities, Dr. Sastry said. It also is important that "we grow new manufacturing approaches" to make new chemistries in a "manufacturable and cost-effective way," she said.

Panel III

Strengthening the Supply Chain

Moderator:
Jim Greenberger
National Alliance for Advanced Technology Batteries

"If anybody had told us in 2008 that the federal government was going to put over $2 billion into advanced batteries and electric drive trains, we would have told that person that he is crazy," said Mr. Greenberger, executive director of the National Alliance for Advanced Technology Batteries (NAATBatt). "It is hard to remember that we lived in a different world just two-and-a-half years ago when it came to advanced battery manufacturing."

NAATBatt, Mr. Greenberger explained, is a trade association of battery manufacturers and other companies engaged in the advanced-battery supply chain. It was founded in 2008 to try to find a way to begin domestic manufacturing of lithium-ion batteries in the United States, he said. "Our mission is to grow the market for advanced electrochemical energy storage in North America."

At the time NAATBatt was founded, America's political leadership did not see petroleum dependence as being a problem, Mr. Greenberger recalled. "Certainly it was not a problem to be solved in this generation and certainly not by that particular Administration." The alliance focuses on affordability, Mr. Greenberger said. "We have got to find ways to bring the cost of batteries down."

The "light of lithium ion remained alive" in the U.S. "only through the work of some hardworking and very talented professionals" in the DOE and private industry, Mr. Greenberg said. Many of those entrepreneurs, he said, "have moved around a bit in the last 10 years."

Now there is a "great, once-in-a-lifetime opportunity," Mr. Greenberger said. "But it also is a tremendous responsibility. It is a responsibility of every one in this room to build an industry that is truly sustainable, to create jobs that are sustainable, and to make some real progress on moving our country away from petroleum dependence."

This panel featured three experts on battery manufacturing and the related supply chain, Mr. Greenberger said. "They know something about what it is going to take to move our industry forward to create a sustainable industry and viable supply chain," he said.

The first speaker, Tom Watson, is vice-president of technology for Johnson Controls Power Solutions and runs the business accelerator program, Mr. Greenberger noted. He hails from Wisconsin.

The next speaker, Mike Reed, joined Magna E-Car Systems in 2009 as general manager of the battery divisions, Mr. Greenberger noted. He started up Magna's North American lithium-ion battery cell and pack manufacturing facility. Mr. Reed has more than 40 years of experience in the battery industry, he explained, with various technical and managerial roles at battery companies. He holds a degree in chemical engineering from Purdue and an MBA from Indiana University "and, I'm sure, some honorary degree from some Michigan university," Mr. Greenberger said.

Building a supply chain for an advanced battery industry isn't only about finished goods, Mr. Greenberger said. "You also are talking about the bottom end of that supply chain, the raw materials," he said.

The next speaker, Linda Gaines of Argonne National Laboratory, addresses the raw materials. Dr. Gaines is a systems analyst at Argonne for transportation research, Mr. Greenberger explained. She holds a bachelor's degree in chemistry and physics from Harvard and a Ph. D. from Columbia University. She began her 30 years of experience at Argonne by writing a series of handbooks on energy and material flows, petroleum refining, organic chemicals, and copper industries that provided background for studies of recycling packaging for discarded tires and other energy-intensive materials, he noted. Dr. Gaines' most recent work has been on reducing petroleum use and other impacts in transportation by recycling batteries and reducing idling.

BATTERY MANUFACTURER PERSPECTIVE

Tom Watson
Johnson Controls

Mr. Watson began by noting that he spent one year in Wisconsin, but 25 years in Ann Arbor. "So I still consider myself a Michigander," he said.

Johnson Controls is involved "in every aspect of energy efficiency," Mr. Watson explained, including devices for homes, workplaces, and autos. It also is committed to corporate sustainability, he added. "This is more than just the economic part of the corporate equation," he said.

Although it must produce revenue and profits, Johnson Controls is committed to social responsibility, "which helps bring jobs to our local areas around the globe as well as support and build the industry." The other part of the company's sustainability model is environmental stewardship, Mr. Watson said. "That is where it touches on all three pieces of our company business," he said. "We deliver greater fuel efficiency and lower emissions in our vehicles and reduce energy consumption in our buildings." Johnson Controls uses a "triple bottom line" model, he said, "as opposed to the single bottom line."

Johnson Controls-Saft, the company's lithium-ion battery joint venture, was first to the market with a lithium-ion application for a mild hybrid vehicle, Mr.

Watson said. It began producing the 19-kilowatt S400 Hybrid battery system for Daimler in March 2009. The same system is used in the BMW 7 series.

Mr. Watson described the electric-vehicle battery supply chain as a "circle of life." Johnson Controls must leverage its partners in materials supply, research-and-development expertise, and infrastructure for charging vehicles, he explained. Internally, it must leverage its own resources in terms of advanced manufacturing capability and energy-efficiency expertise in all parts the company for the lithium-ion business, he said. Because Johnson Controls is a large corporation, it also has the ability to service product warranties, "something our customers are extremely interested in," he said.

Johnson Controls received a sizable grant through the Recovery Act to build a manufacturing plant in Michigan, he noted. "However, that won't be sustained if we don't continuously have new innovations that will drive higher levels of energy and power density and more cost-effectiveness in the products made in that plant."

To keep innovating, Johnson Controls is collaborating with Argonne and Oak Ridge national laboratories and many universities, Mr. Watson said. Existing and future suppliers are working with the company on new materials. Johnson Controls also works with technology start-ups. "We can't ignore the ability of some small, fast-moving innovators to come up with innovations even for a large company like ours," he said.

Johnson Controls has assembled a "world-class team" of suppliers and manufacturers for its supply chain, Mr. Watson said. "We have gone around the world to find what we believe are the best suppliers," he said. The supply chain extends all the way to recycling. In addition to Toxco and Umicore, which are major recyclers, Johnson Controls itself is the world's largest recycler of lead-acid batteries, Mr. Watson explained. It is leveraging this expertise in processing used batteries and recovering and reusing materials for its lithium-ion business.

One problem is that most key suppliers are based overseas, Mr. Watson said. For Johnson Controls' U.S.-based manufacturing plants, "we really like to work with local suppliers," he said. "It follows along with the social responsibility of our Triple Bottom Line." But the cells, separators, and cathode materials "pretty much are coming out of Europe, Japan, and Korea," Mr. Watson explained. There is more U.S. involvement with battery packs in terms of software and mechanical components, Mr. Watson said. "But again, a lot of the supply base is offshore.

Some suppliers are developing capability in the U.S. Johnson Controls also is requiring foreign suppliers to set up manufacturing in the U.S., Mr. Watson said. "We need to do more in terms of capturing a larger part of the value chain

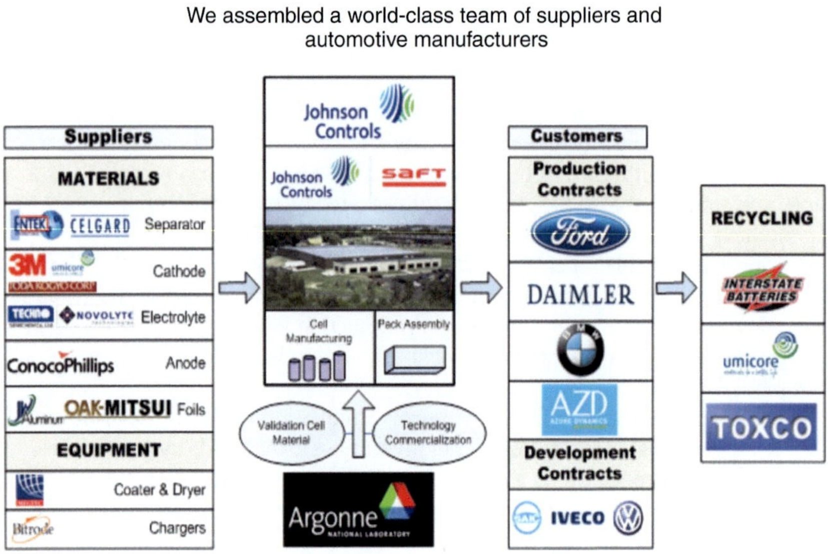

FIGURE 8 Johnson Controls' supply chain.
SOURCE: Tom Watson, Presentation at July 26-27, 2010 National Academies Symposium on "Building the U.S. Battery Industry for Electric Drive Vehicles: Progress, Challenges, and Opportunities."

in the U.S.," he said. "This is something we need to remedy not only with manufacturing but also with R&D.

Mr. Watson concluded with several quotes from Alexander Hamilton to show that the importance of manufacturing was recognized even at the foundations of the nation:

> "*Not only the wealth, but the independence and security of a country, appear to be materially connected with the prosperity of manufacturers. Every nation...ought to endeavor to possess within itself all the essentials of a national supply. These comprise the means of subsistence, habitation, clothing, and defense.*"[20]

[20] Alexander Hamilton, "Report to Congress on the Subject of Manufacturers," Dec. 5, 1791, published in the Annals of the Second Congress, 1793.

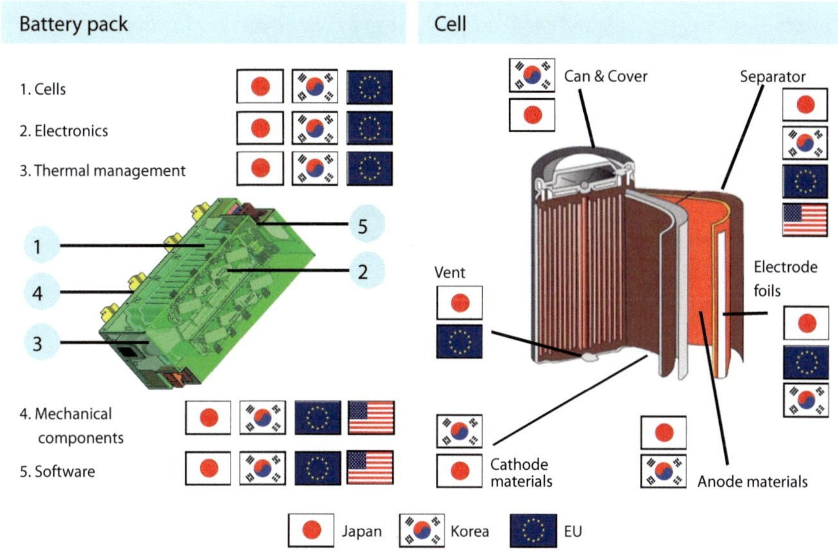

FIGURE 9 Most of the key supply base is in foreign countries.
SOURCE: Tom Watson, Presentation at July 26-27, 2010 National Academies Symposium on "Building the U.S. Battery Industry for Electric Drive Vehicles: Progress, Challenges, and Opportunities."

DEFINING THE SUPPLY CHAIN: GAPS AND OPPORTUNITIES

Michael E. Reed
Magna E-Car Systems

As one of the world's largest tier-one suppliers to automobile manufacturers, Magna Steyr buys lithium ion cells from numerous companies. "We are in the middle of the supply chain," explained Mr. Reed, who runs Magna E-Car Systems, the company's lithium ion cell and battery pack division.

Magna is one of the largest, most diversified manufacturers in North America, Mr. Reed said. Based in Aurora, Ontario, the $17.6 billion company has 120 facilities in the U.S. and Canada and around 74,000 employees in 25 countries.

Magna Steyr divisions have been involved in electric vehicles for many years, Mr. Reed said. Magna Electronics, for example, makes charging and electric controls for hybrid and electric vehicles. Magna has been developing lithium-ion batteries for more than six years, especially in Europe. The Magna Cosma division makes electric vehicle structures. Magna Powertrain produces everything from axels and start-stop systems to water pumps for electric vehicles.

Magna E-Car Systems was started in 2009 and has its headquarters in Auburn Hills, Mich. The unit is "the face to the customer" for all Magna products and services used in electric-vehicle programs of OEMs, Mr. Reed explained. "This is a new, focused effort by Magna to aggressively grow the whole vehicle electrification business," he said. "Our chairman and founder, Frank Stronach, believes this can be as big as Magna itself in the next few years as our business evolves."

Magna offers the gamut of services and products to automakers. It integrates components into vehicle systems, develops complete vehicle solutions, and designs and builds cars on a turnkey basis. In Auburn Hills, Magna E-Car has facilities for battery materials testing and builds hybrid, plug-in hybrid, and full-battery electric systems.

Magna E-Cars has a good view of the entire lithium-ion battery supply chain. It buys cells from various manufactures that it uses to make battery packs. "We probably have benchmarked about every available technology and continue to do that," Mr. Reed said. It also is a buyer of technology and services to develop battery packs and the capital equipment for making batteries, as well as all of the materials and components

Having the supply base scattered across the globe, mainly in Asia presents, complicates logistics. "Obviously, this creates communications issues, due to both time zones and languages," Mr. Reed said. "That is a challenge for us dealing with this particular type of technology."

Transportation and customs-clearance also are serious issues. "Moving people and hazardous materials across borders is challenging," Mr. Reed said. That is especially true with lithium-ion battery materials, which present safety concerns that are still evolving. Countries set a wide range of safety and material regulations for materials, machinery, and the products themselves. "The complexity of the supply chain obviously adds significant costs," Mr. Reed said. Companies also must carry substantial inventories "to protect your operations when this very long and diverse supply chain is part of your business."

There are several initiatives to invest in North American production of cells and materials. "But that is still a far cry short of the full supply chain that needs to be put into place," Mr. Reed said. Conductive materials, foils, separators, and electrolytes all primarily come from Asia, he said, "although there are many potential qualified suppliers here in North America." One significant material for a pouch sell, the laminate, "is almost exclusively confined to Japan at the moment," he said.

One major advantage of big Asian lithium-ion battery companies is that they are vertically integrated or control their supply chains, Mr. Reed said. The supply chain was established essentially to support the electronics industry in Japan, he said. That provides major cost benefits to Japanese manufacturers and restricts access to key parts and materials. While still largely controlled by Japan, the supply industry is expanding to South Korea and China. "But we have a lot of catch-up to do to become a viable competitor in this market," he said.

Investment in North America in lithium-ion cell production in the past year has been "impressive," Mr. Reed said. "But I don't think it has been balanced by necessary investment in the supply chain itself," he said. The uncertainty of market volumes and timing are causing many delays in investment, and government incentives have been "relatively short term," he said. "In terms of the business cycle most battery manufacturers operate on, we need longer-term vision—five to seven years or longer—to make the incentives being considered more effective."

Small volumes are another problem. Most electrified vehicle programs by automakers involve a few thousand vehicles a year. "Very few people are announcing programs in the tens of thousands per year or higher," Mr. Reed pointed out. The costs of developing and validating products and applications for so many small programs "is really prohibitive in the way the industry is evolving at this point," he said.

Companies are still investing in these areas, often thanks to federal incentives that cover "50 cents on the dollar," Mr. Reed. "Still, big investments are required to make sure these products work, that they have a long-term performance capability, that they will meet customers' requirements, and that they will not bankrupt a company because of any warranty exposure."

The technologies involved also make the validation process long and expensive. Over the last 20 years, lithium-ion cell have learned a lot. "But they also have developed a product that is not suited for the automotive industry," he said. "It is one thing to produce a cell that lasts two or three years in a laptop or cell phone before you turn it in for your next model. But people don't turn in their vehicles that often."

Government and industry could help by promoting standardization and testing, Mr. Reed said. Current testing standards for batteries are evolving and not fully developed. "Literally every OEM customer has its own special set of requirements that drives this whole process," he said. "Often, you have cell or pack technology that has been developed and qualified to one set of standards, but you may still need to spend millions of dollars to re-qualify it for another OEM's specific set of standards."

There has been a lot of discussion of standardizing cell sizes, such as in Germany, Mr. Reed noted. His experience in the battery industry, however, "suggests that this is something that is not going to happen by committee," he said. "It's going to happen by success in the marketplace, and the winners will set the standards that will then help lower the costs long-term. It is going to require volume to get those answers."

It also would help if producers of cells and components develop "truly automotive-grade hardware with well-known reliability and life-performance characteristics," Mr. Reed said. That would enable car makers to engineer those components into their future vehicles. "Much of that does not exist today," he said. "It's not just a cell-development issue. It is development of the many components that go into the battery pack."

Continued investment is needed in advanced storage technology, Mr. Reed said. "But unlike in the past, when investment focused on the 10-year-plus time horizon, we need to be focusing on the development side of R&D to get us faster into applications so that we have products to sell."

Government programs to boost demand also are very important, he said. "To pick winners by putting grant and loan guarantees into selective companies, and to pick winners too early, could end up being counter-productive if those companies cannot sustain themselves with volume," Mr. Reed said. Policies that promote demand, develop a diverse supply base, and let companies compete and succeed are desirable, he said. "Anything we can do to increase volumes of hybrid and electric vehicles and reduce the uncertainty over when and how large this business is going to be will help move this business forward."

BATTERY MATERIALS AVAILABILITY AND RECYCLING

Linda Gaines
Argonne National Laboratory

The main motivation for the transportation electrification push is easing America's dependence on imported petroleum said Dr. Gaines of Argonne's Center for Transportation Research. But there also have been concerns about American dependence on an essential ingredient for advanced batteries—lithium. "The main motivation of this talk is to make sure we are not going to run out of something important," she said. "There was a big scare produced by one individual, actually, that we might be running out of lithium for lithium ion batteries."

David Howell and others at the Energy Department asked her to look into the issue, Dr. Gaines explained. Another part of her work is to look at the environmental impact of recycling batteries, she said.

Argonne tried to address the question of lithium supply "in the most logical way possible," Dr. Gaines said. It asked how many electric cars there will be, how much lithium they will use, and how that compares to how much lithium there is.

The study used scenarios prepared for the DOE by Argonne's Washington staff. One scenario was "business as usual." Another assumed hydrogen vehicles would be successful. There also was a "maximum electric" scenario, Dr. Gaines said, which projected how many electric vehicles of different types would be on the market at different time scales.

Under the "maximum electric" scenario, hybrids would soon account for 25 percent of the U.S. car market by around 2020 and stay at that level through 2040. Demand for plug-in hybrids would take off around 2020 and take about 60 percent of U.S. car market in 2050. Full-battery electrics would become a factor after 2020 but still account for just 10 percent of the market in 2050.[21] Dr. Gaines said she suspects the scenarios may be "a little pessimistic" in terms of when hybrids and plug-ins will penetrate the market. Even if those sales levels are achieved sooner, however, "it doesn't change the basic numbers," she said.

The next step was to calculate how much lithium ion would be needed in the batteries. "Luckily, Argonne has a really excellent battery-development group," she said. Co-author Paul Nelson helped produce careful calculations for different battery chemistries for vehicles of different ranges.

The study looked at four different chemistries, all of which used lithium in the cathodes. They are nickel-cobalt-aluminum oxide (NCA), lithium iron phosphate (LFP), and two lithium manganese dioxide types, one (LMO) with a graphite anode and another (LMO-TiO) with a lithium-titanium based anode. The LMO-TiO was the only chemistry of the four using lithium in its anode. The others use graphite.

For vehicles with a 100-mile range, the most lithium required in a battery was 12.7 kilograms, Dr. Gaines said. That was for the battery using the LMO-TiO chemistry. "That is a fair amount, but most batteries use a lot less," she explained. The NCA battery uses 7.4 kilograms in a car with a 100-mile range. The LMO version requires just 3.4 kilograms.

The batteries using the lithium-titanate anode, however, are unlikely to catch on for vehicles with a 100-mile range because they would weigh 500 kilograms. . Dr. Gaines said. "I can't imagine a lot of people driving around with 1,000 pounds of battery," she said.

Based on these assumptions, 50,000 to 60,000 tons of lithium would be used in electric cars on American roads by 2050, Dr. Gaines said. Currently, less than 30,000 tons of lithium is produced in the world each year, but that is rising rapidly. It won't be until around 2030 that U.S. lithium demand would match current production levels, even in the rapid penetration scenario.

Lithium production can be increased. Dr. Gaines noted that she recently attended a conference in Las Vegas called Lithium Supply and Markets 2010. "If you believe the people who were there, there are at least 100 companies exploring for lithium at 150 different sites. If they all produce what they expect to be producing, the production over the next 20 to 30 years could be way more than even the most optimistic demand scenarios," she said.

Dr. Gaines cautioned that the demand projections are just based on scenarios. "They are not meant to be what we expect to happen," she said. "It is a 'what if,' and is very optimistic." It was an exercise to determine how much material would possibly be needed if "these vehicles come as quickly as we imagine," she explained.

[21] Phil Peterson, Margaret Singh, Steve Plotkin, and Jim Moore, "Multipath Transportation Futures Study: Results from Phase 1," Office of Energy Efficiency and Renewable Energy, U.S. Department of Energy, March 9, 2007 (http://www1.eere.energy.gov/ba/pba/pdfs/multipath_ppt.pdf.

The team also looked at how much lithium would be needed it were recycled. More than 40,000 tons of the material used in 2050 could be recycled, the study concluded. As a result, the "net virgin material needed" for the U.S. electric car market would peak at around 25,000 tons a year in the 2030s and decline to less than 15,000 tons by 2050—below current world production. "Recycling drastically reduces the amount of virgin material that you would need," she said. "We certainly know that recycling is feasible" because it is done with lead-acid batteries.

The re-use of car batteries at the end of their life would also affect lithium demand. While those batteries may not be good enough to operate a car, they may be fine for utility storage, she said. Re-use would raise the amount of lithium needed by about 10,000 tons annually because it would delay the material recovered through recycling.

All of these estimates apply only to U.S. consumption, however. Large volumes of lithium-ion batteries also will be required in the rest of the world, Dr. Gaines noted. The Argonne team used estimates from the International Energy Agency that an assumption "even more aggressive than what we did," she said.

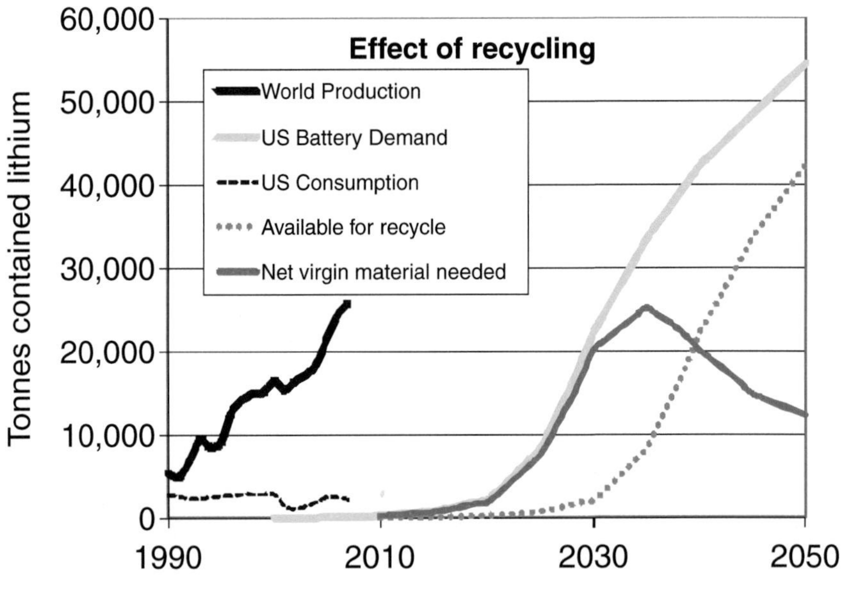

FIGURE 10 Recycling can drastically reduce virgin Lithium demand.
SOURCE: Linda Gaines, Presentation at July 26-27, 2010 National Academies Symposium on "Building the U.S. Battery Industry for Electric Drive Vehicles: Progress, Challenges, and Opportunities."

It assumed the target of reducing carbon by 80 percent would actually be accomplished, "largely through rapid introduction of electric vehicles in the world market," she said. Dr. Gaines called that an "incredibly uncertain scenario."

Developing global estimates is complicated because lithium demand is influenced by the size and range of the vehicle, Ms. Gaines explained. The IEA estimates assume cars will have batteries of 12 to 18 kilowatt hours of storage capacity, "which I think is rather large for the rest of the world," she said. Electric vehicles also could be bicycles, she pointed out.

If these projections were true, however, world lithium demand would reach about 450,000 metric tons by 2050, Dr. Gaines said. If the average battery size of these vehicles is much smaller, demand would drop to around 200,000. And if these smaller batteries were recycled, lithium demand would drop to around 100,000 tons. In other words, lithium demand would grow to four times current production compared to 20 times if the more aggressive IEA assumptions are used. "It is not unreasonable to assume that you can increase current world production by a factor of four in 40 years," she said.

Batteries account for one-quarter of global lithium use, and that share will keep rising, Dr. Gaines predicted. She also noted that use of lithium-ion batteries for cars is growing much faster than for electronics products.

When one looks even at conservative estimates of known lithium reserves in the world, however, "one sees that we are not about to use up all the material in the ground," Dr. Gaines said. Cumulative demand for lithium until 2050 is estimated to be 6.5 million metric tons if large car batteries are used and there is no recycling, she said. That would drop to 2 million metric tons if smaller batteries are used and there is recycling.

The U.S. Geologic Survey estimates reserves in mines around the world at 9.9 million metric tons, with 7.5 million metric tons in Chile. The USGS puts known world reserves at 25.5 million metric tons.[22] Other estimates are higher. In fact, U.S. demand could be met by domestic lithium resources in Nevada and California if they are developed and recycling is implemented, Dr. Gaines said.

There are concerns with materials besides lithium. Dr. Gaines noted that U.S. demand for cobalt and possibly nickel would make a serious dent in global reserves by 2050. "But I don't suspect we will be using that much cobalt in batteries," she said. "I suspect we will be moving away, as manufacturers already are, from cobalt as a significant component of the cathode material." Demand for aluminum, phosphate, manganese, and titanium for car batteries would account for a far less significant share of global reserves.

The conclusion from these data is that "lithium ion batteries can at least give us a bridge to the future until something better comes along," Dr. Gaines said. "I think it could be a fairly long term."

---

[22] Data: U.S. Geological Survey, revised January 2010 data. See (http://minerals.usgs.gov/minerals/pubs/commodity/lithium/mcs-2010-lithi.pdf). These numbers were revised again in 2011.

Many warnings over the years "that we are running out of something have been grossly wrong," Dr. Gaines observed. The Club of Rome, for example, famously predicted in 1972 that the world would "run out of gold in 1981, mercury and silver by 1985, tin by 1987, and petroleum, copper, lead, and natural gas by 1992."[23] "So I don't think we should be too alarmist," she said.

In terms of recycling, "a pet project of mine," different processes are currently being developed, Dr. Gaines said. "The interesting thing about them is that they recover material at different stages of the life cycle," she said. Some processes go all the way to smelters. At the other extreme, processes try to recover lithium as battery-grade material that can quickly be put back into batteries, she said.

Smelting recovery processes are working now, Dr. Gaines said. "You can throw anything in and get cobalt, nickel, or whatever out the bottom," she said. Lithium, however, goes into the slag. "It could be recovered, but it is not," she said.

Toxco, a company that received a DOE grant, will be recovering lithium in a new plant in Ohio, Dr. Gaines noted. Another company, OnTo Technologies "has demonstrated it can take a uniform stream of batteries, break them up, recover the electrolyte, recover the cathode and anode material, and re-use the materials to make a new battery and have it perform reasonably well," she said. "It is a low-energy process, you don't need high temperature, and they do believe they can get battery-grade material back out."

Argonne would like to now determine which of these processes "makes the most sense from an energy and environmental standpoint," Dr. Gaines said. "Stay tuned for that."

A new process being developed at Argonne, meanwhile, attempts to take waste plastics bags (and could use plastics from batteries) and turn them into carbon nano-tubes, Dr. Gaines said. "That would be a way to get that material back into batteries or up-cycle it into something even more valuable," she said.

DISCUSSION

Mr. Greenberger said he was intrigued by Mr. Watson's point about Johnson Controls' foreign suppliers. "My quick count was that it seems about 80 percent of the materials you are buying for cells and components are not sourced in any major quantity in the United States," he said. He asked Mr. Watson to elaborate and suggest what North America must do to expand its supply base for those components.

Materials that Johnson Controls uses to make cells in France come from Asia, Mr. Watson noted. "By and large, as we started to set up shop in North America, we found a lack of a supply base here in the U.S.," he said.

---

[23] Donella H. Meadows et al., *The Limits to Growth*, Universe Books; 2nd edition (February 18, 1974) The authors note that their purpose was not to make specific predictions, but to explore how exponential growth interacts with finite resources.

As a result, Johnson Controls has required each of its materials suppliers to build processing factories in the U.S. Each factory or supplier has a "variable degree of vertical integration" in the U.S. versus what they are doing overseas, he said. "We would really like to encourage a great mix of vertical integration in the U.S."

One way to encourage that is to stimulate a strong R&D push "to get the new emerging suppliers as well as the established suppliers to be conducting their R&D onshore here," Mr. Watson said.

Mr. Reed of Magna noted that Japanese suppliers are vertically integrated within their customer base as well. In consumer electronics, "they really got into batteries as a means to help sell what they have," he said. "That means companies that are not a part of that *keiretsu* don't get the same access to new R&D coming out of their labs as do those who are part of the *keiretsu*. So to be able to establish an R&D base in the U.S. at the material supply level will allow battery makers in the U.S. to do more collaboration with suppliers and compete more effectively."

Mr. Reed said the biggest thing that can be done is to expand the U.S. market for electric vehicles. Generally, cells and battery packs for North American will be sourced in North America. Those sold in Europe will be sourced in Europe, and those sold in Asia will be sourced in Asia. "If you look at the history of the supply chain and the number of factories, I think it is unlikely any one country is going to dominate and export to the rest of the world over the long term," he said. "I also think it is unlikely the U.S. is going to become the battery supplier to the world. So I think the faster we can grow the market, whether for automobiles or commercial vehicles, the faster we will drive the supply base that is necessary."

Restoring U.S. lithium production would help, Dr. Gaines said. The U.S. produced lithium until cheaper South American sources became available, she said. "At least we will have raw material here," she said. "Obviously, the battery manufacturers need the lithium compounds to make batteries."

Dr. Gaines said it also seems feasible for some U.S. chemical companies to make lithium carbonate and metal oxides. "It seems unnecessary to send lithium carbonate made in the U.S. and shipped to Asia to be made into cathode materials that then are shipped back here to make batteries," Dr. Gaines said.

Mr. Greenberger asked how much of an issue standardization is for the industry. "How much cost can we squeeze out of the process by standardization?" he asked. "And if it's not going to happen by committee, how is it going to happen?"

It is good to many different developers of materials, cell designs, pack technology, and vehicles, Mr. Reed responded. "I think diversity is going to allow the winners to be developed and emerge," he said. The first step is to standardize the way materials and cells are judged, Mr. Reed said. That way, potential suppliers "have a clear understanding of what the expectation is." If expectations are fairly consistent across customers, then the costs of qualification and development "can be kept to a reasonable."

Most of the DOE budget for batteries goes to development, Mr. Reed noted. "What people often lose sight of is that once you have a material, cell, or product, you then have to go through a full series of qualifications to get them into an OEM program," he said. "It may cost you $2 million to $3 million to take a fully engineered cell through a full range of qualifications for one or more customers." That process for a battery pack can take $10 million after it's been developed all the way to make sure it has the life and durability to make sure it is a successful product, he said.

Still, Mr. Reed said he doesn't expect standard products to emerge until high volumes of vehicles are produced. That will determine the "winners in the survival-of-the-fittest process," he said. "Then I think we will have standards cells. But that is many years into the future."

Mr. Watson said he does not see a rush toward standard cell sizes or capacity. Whether the customer is an automaker or consumer-electronics company, "I'm not sure that we as a battery maker want to dictate the way they do their businesses," he said. Standardization could help testing, charging, handling, and transportation standards, he said.

Panel IV

Market Drivers: Creating Demand for Electric Vehicles

Moderator:
Robert Kruse
EV Consulting LLC

This panel addresses what is required to create market demand for electric vehicles, explained moderator Robert Kruse, the founding principal of EV Consulting LLC and former executive director of global vehicle engineering for hybrids, electric vehicles and batteries at General Motors. "I have a pretty impressive panel to talk about various aspects of what can be done to spur electric-vehicle adoption," he said.

The first topic addressed market incentives. Speaker Daniel Spieling is a professor of civil engineering and environmental science and policy at the University of California at Davis, Mr. Kruse noted. He is founding director of the Institute of Transportation Studies and acting director of the Energy Efficiency Center, both at UC-Davis.

Dr. Sperling also holds an automotive engineering seat on the California Air Resources Board that has oversight responsibilities over the state's policies regarding climate-change, alternative fuels, vehicle travel, land use, and the zero-emission vehicles program. He is co-director of the California Low-Carbon Fuel Study[24] and chairs the Future Mobility Council of the Davis World Economic Forum. Dr. Sperling is an active member of 13 National Academy committees, Mr. Kruse noted.

The next speaker, Gary Smyth, was there to provide "an industry perspective in transforming the auto industry," Mr. Kruse said. Mr. Smyth is executive director of North American research-and-development labs for General Motors. He has "a notable 20-year career with General Motors," he explained, primarily in advanced power trains with early work on advanced

---

[24] Researchers at the University of California-Davis and University of California-Berkeley have been conducting a study of California's Low Carbon Fuel Standard program since 2007. The first part of the study, Alexander E. Farrell and Daniel Sperling, "A Low-Carbon Fuel Standard for California, Part 1: Technical Analysis," Institute of Transportation Studies, University of California, Davis, May 2007, Research Report UCD-ITS-RR-07-07. Also see Alexander E. Farrell and Daniel Sperling. "A Low-Carbon Fuel Standard for California, Part 2: A Policy Analysis, Institute of Transportation Studies, University of California, Davis, August 2007, Research Report UCD-ITS-RR-07-08.

hybrid cylinder de-activation and direct fuel-injection systems. Mr. Smyth earned a bachelor's and Ph. D. degree from Queens University in Belfast.

The final speaker, Bill Van Amburg, discussed early adoption of hybrid vehicles. Mr. Van Amburg is senior vice-president of CALSTART, which Mr. Greenberger described as "a non-profit, fuel–neutral, membership-supported consortium." He overseas five programs: heavy hybrids, new fuels, technology commercialization, fleet analysis, and consulting and industry services. "He brings 25 years of experience in marketing and technology commercialization," Mr. Greenberger said. Mr. Van Amburg's academic credentials include degrees from the UCLA Anderson School for Management, Stanford, and the University of California-Berkeley.

INCENTIVES FOR THE ELECTRIC VEHICLE MARKET

Daniel Sperling
University of California-Davis

Because he hails from "that other land, the Left Coast, the foreign country of California," Dr. Sperling began, "I am going to have a little different perspective here today."

His presentation aimed to achieve four missions, Dr. Sperling said. They were to emphasize the important role of universities, explain climate and energy policy from the California perspective, promote his book *Two Billion Cars*,[25] and explain "the market for electric vehicles and what that means for battery design, which is the main reason we are here." His focus, he said, is to address what consumers really want in electric vehicles.

Research by his institute on American consumers has revealed some interesting insights, Dr. Sperling said. "What American consumers seem to want in electric vehicles is quite different from what I have been hearing today" from speakers in the symposium.

The success of plug-in electric vehicles will depend largely on government policy, advances in battery technology, fuel prices, and consumer response to products, Dr. Sperling explained. "There are a lot of ways for companies to fall into the Valley of Death," he said. "But a lot of it has to do with misjudging and misunderstanding consumer responses and consumer behavior."

State and local governments have implemented plenty of policies aimed at supporting electric vehicles, Dr. Sperling pointed out. The main policies include zero-emission standards by several states, greenhouse-gas emission and fuel standards for vehicles "that have very aggressive incentives for electric vehicles," tax credits for consumer purchases of low-carbon vehicles, subsidies

---

[25] Daniel Sperling and Deborah Gordon, *Two Billion Cars: Driving Toward Sustainability,* New York: Oxford University Press USA, 2009.

for manufacturers, and government-sponsored research and development, he said. While "this mélange of policies" does not always seem consistent, he said, they are "consistent in their support of electric vehicles."

New Environmental Protection Agency standards for greenhouse gas emissions are especially important, Dr. Sperling said. Standards already being put into effect in California and that the Obama Administration plans to adopt for the entire nation call for emission reductions of 30 percent to 40 percent by 2016.[26] "That is aggressive," he said.

California is the process of issuing "the next level of standards," Dr. Sperling explained. "We are talking about at least a 3 percent reduction in greenhouse gases per mile a year, starting in 2017._[27] So however aggressive those 2016 numbers are, it is going to get even more aggressive after that." Dr. Sperling said California is working very closely with the EPA and hopes the rest of the U.S. will quickly follow its lead.

In addition to putting a lot of pressure on automakers to improve fuel-efficiency, these standards are designed to push electric vehicles, Dr. Sperling explained. For example, electric vehicles will be rated as if they emit zero grams of greenhouse gasses per mile "even though in reality their lifecycle emissions are much more than that," he observed. "The intent is to incentivize EVs. There is tremendous pressure to move toward electric-drive technologies. This requirement is going to make that even more forceful."

California and 10 other states have required that a certain percentage of cars sold in the state be zero-emission vehicles.[28] California requires major automakers to make available 25,000 such vehicles in the state by 2014 and 50,000 by 2017. "And we are going to adopt numbers that are far more aggressive than that in 2018 and beyond," Dr. Sperling said.

Low-carbon fuel standards adopted by California are another source of pressure. The state requires a 10 percent reduction in carbon-intensity of all fuels, whether they are natural gas, petroleum, bio-fuels, or hydrogen. "What this does is require that the carbon content of fuels be steadily reduced over time," he explained. The greenhouse gas benefits of bio-fuels, especially those derived from foods, are modest, Dr. Sperling noted. "So again, on the fuel side we will see strong incentive to move toward electricity as a transportation fuel." Local governments in California also are promoting the electric-vehicle industry. Municipalities are setting targets for reducing carbon from transportation.

---

[26] The U.S. Environmental Protection Agency and the Department of Transportation's National Highway Traffic Safety Administration (NHTSA) are finalizing greenhouse gas-emission standards for model years 2012 to 2016 under the Energy Policy and Conservation Act. For details, see *<http://www.epa.gov/oms/climate/regulations/420f10014.htm>*.

[27] In July 2011, the White House, together with most of the major car companies, announced a proposal to reduce car emissions (and fuel consumption) by 5 percent per year from 2017 to 2025, and light trucks by 3.5 percent per year. These new standards are scheduled to be adopted by California in January 2012 and EPA and DOT later in 2012. see **Federal Register** / Vol. 76, No. 153 / Tuesday, August 9, 2011 / Proposed Rules.

[28] A zero-emissions vehicle, or ZEV, emits no tailpipe emissions from the onboard source of power.

In terms of the consumer market, Dr. Sperling explained that his Institute of Transportation Studies at UC Davis has been studying alternatives fuels, from methanol to hydrogen, for 25 years. "We have done them all. We have done lots of studies and worked with lots of car companies," he said.

Although Dr. Sperling said he is optimistic about the future of electric-drive vehicles, he added a cautionary note. It has taken hybrids 10 years to reach a market penetration rate of 3 percent "with a technology that is cheaper than plug-in hybrids and battery electrics, and that doesn't require any change in consumer behavior nor change in infrastructure," he pointed out. It is important to keep this conservative reality of markets in mind, he added.

On the other hand, much of the pessimism one hears about electric vehicles—including at this meeting—is based on consumer assumptions that may not be valid. The dominant way of thinking of electric vehicles is that they need super-advanced batteries to make them more like gas-powered vehicles, with similar driving ranges and longer recharge times, he said. That implies "we need public recharging infrastructure so people can recharge whenever they get range anxiety," Dr. Sperling said. "This is the way engineering experts think about it."

A different way to look at the issue is to study what really motivates consumers. "Electric vehicles give access to a whole new set of values and benefits," Dr. Sperling said. "Now you can avoid gas stations. Plus, everyone who drives EVs loves the driving feel of an electric vehicle." Another appeal, the Institute's surveys have found, is that "driving an EV means not financing terrorists, shorthand for not having to import oil. They don't have to support Big Oil, nor wars in the Middle East. They can support energy independence and reduce climate change, air pollution, and noise. There are many good things about electric vehicles that resonate with people."

Research also is finding that "people are remarkably willing to adapt to changing conditions and constraints if they see some value in doing so," Dr. Sperling said. He cited research his Institute conducted with BMW to evaluate consumer experience with Mini E cars. The Institute conducted intense household interviews of Mini E drivers that "used the vehicles day in and day out, so they have considerable experience with them," he explained.

The study found that around one-third of Mini E drivers "are perfectly comfortable with a 100-mile range as long as they have home base charging" he said, while about half of drivers say this range "kind of pushes them," Dr. Sperling said. But with minimal adaptations, even the 100-mile range will work for them."

After using the vehicle for some time, only about one-sixth of Mini E drivers said the 100-mile range "was really problematic for them," he said. "But in this case, many were willing to engage in different planning and adaptation to make it work because they really like the idea of having an electric vehicle."

Many drivers learned to cope with the limitations. In Los Angeles, for example, drivers began plotting out driving distances on Google to determine whether they would be able to make it. They studied whether there were places to charge along the way, or whether they could swap vehicles with others in the family for that day. "It turned out that a lot of them said it was fun," he said. "It was a positive experience. It was a game. It was competitive."

The studies in Los Angeles also found that most Mini E drivers charged their vehicles at home at night. "Even those who had charging at their work place didn't really need it," Dr. Sperling said. "They thought they didn't even want to ask their employers about it and whether they had to pay for it. "They didn't even want to ask their employers about it and whether they had to pay for it. They didn't even want to deal with it. What we see are these people having a sense of independence."

These findings are pertinent to the question of public charging infrastructure, Dr. Sperling said. The federal government and California are investing a lot in public charging stations. "The dominate way of thinking is that, yes, we need lots of public charging stations to remove this range anxiety," he said. "But what we are finding in all our research and experience is that people don't use public charging, even when it is available."

Dr. Sperling pointed to the experience of Tokyo. The public power utility urged people to buy electric cars but generated little consumer response. So it set up public charging stations. This did indeed generate a response, and people started buying electric cars. The stations, however, were rarely used. "What is going on here is that the public charging stations have psychological value, but people don't use them," he said. "The question is what we do with this information. There is no business model there because it won't be used very much." One lesson is that there is no business model, there because drivers don't use public charging very much." It also means that a minimal number of stations are needed, at least in the beginning, to address consumer anxiety over the range of their cars. "Providing public charging does not appear to be critical to building an electric vehicle industry," he said.

The real challenge is that consumers have no real experience with EVs, and thus it is difficult to predict consumer behavior, Dr. Sperling explained. There were a few EVs in the 1990s, such as the EV1 and Rav4 EV. But electric vehicles have never been mass-produced. The Nissan Leaf and GM Volt won't be available until December 2010. "So one conclusion you will see is that we need a lot more research and understanding of consumer behavior, because all the insights I have presented here, based on many studies over many years, are tentative and preliminary," Dr. Sperling said.

The first overall conclusion of the Institute's research is that people like the concept of an electric car, he said. The second as that "the more experience they have, the more comfortable they become with owning an electric car." The general pattern is that most people have very positive impressions of the vehicles before they use them. Their favorable impression then declines the first weeks of using the vehicles as they experience limited range, but as they gain more experience, their positive attitudes return. "Some consumers will never buy the vehicles because of the limited range. They just don't want them," Dr.

Sperling said. "But there are a whole lot of them, if they stick with it and if marketing sticks with it, who will get more interested in owning an electric car."

Research into consumer expectations for electric vehicle batteries also yielded surprising findings. Dr. Sperling noted that various organizations have adopted aggressive targets and design goals for range, electricity consumption, recharge times, and cost. These goals are based on "what experts say we need for consumers to be willing to buy the vehicles," he said.

The Institute conducted a large national study of what kind of hybrid vehicles U.S. consumers actually prefer in terms of cost, range, and recharge time. Most consumers said they were satisfied with batteries that supply only around 2 kilowatt hours. With that kind of battery, consumers were told, they would get around 75 miles per gallon for 10 miles in a vehicle that used a blend of electricity and gasoline. "That is not very much electricity storage," he said. "But that is what people said they wanted, when they compared how much they are willing to pay for how much range and how much fuel economy."

Consumers in the study rated fuel economy and reduced cost very highly as key factors. "It turns out that when you do the analysis of the fuel-economy performance of a PHEV 10 or PHEV 20 vehicle, with their small batteries, one gets a large percent of the benefits that one might get from a PHEV 40 with its much larger battery," Dr. Sperling said. "To the extent this is true, if we think about electric vehicles and plug-in hybrids in terms of their positive attributes rather than their problems, and think about what people are willing to pay for, we conclude that higher-performing batteries are not essential to launch the EV industry," he said. He noted that the findings are tentative, but that he also believes they are robust.

The message is that "you can sell the equivalent of a PHEV 10 or 20, probably fairly easily," Dr. Sperling said. "It looks like this vehicle technology will get the most market penetration and will provide the most benefit in terms of fuel reduction and greenhouse gases. And it gets you on the path toward the bigger battery size of the future and it gets people more comfortable with battery use."

Dr. Sperling noted that a group at General Motors is promoting ultra-small electric vehicles—with small batteries. Such vehicles are more likely to succeed in markets such as China or European cities rather than the U.S., he said. "This approach should be part of the mix as we learn how to work with attributes of batteries and electric vehicles and how to make them successful," he said. "I would suggest that if we continue to follow the path we're on, trying to create an electric vehicle that is analogous to a gasoline vehicle, we are doomed to failure."

The GM Volt and Nissan Leaf "are incredibly important in terms of creating market presence for electric vehicles and early market acceptance," Dr. Sperling said. "But what I am suggesting is that in order to get a viable mass market for electric vehicles, there are different ways of thinking about it that need to be pursued."

THE INDUSTRY PERSPECTIVE: TRANSFORMING THE AUTOMOTIVE INDUSTRY

Gary Smyth
General Motors

Mr. Smyth began by noting that he agreed with much of what Dr. Sperling said. "You really have got to look at what personal transportation in the future will be," he said. "Is it the same as today? Is the current strategy sustainable for the future? Clearly, there will be a lot of changes."

As one conducts research for vehicles such as the Volt, "you really begin to understand that what you need for the mega cities and hyper cities is very different than what you need in Texas and the Midwest," Mr. Smyth said. "It really is about a portfolio of solutions."

When one steps back and looks at "what we are really trying to achieve," Dr. Smyth said, "the objective becomes energy sustainability." "That energy will have to come from multiple sources," he said. "And because of the environment, it has to be low carbon. At the same time, we have to make sure we are developing economic prosperity while we do all of that. That is the challenge we have looking forward."

Dr. Smyth said GM will produce a full line of vehicles using different technologies. "Electrification is an important part of the future, but it is not the only part," he said. "It is certainly necessary, but not sufficient." Therefore, the industry must continue to dramatically improve conventional systems. "The whole portfolio of hybridization and electrification is extremely important to us." GM also must work on technologies such fuel cells and hydrogen fuel cells, he said, where companies are getting in position to introduce commercially viable products by 2015 and 2016. All of these technologies will be important, "whether you look at it from an environmental perspective or energy perspective," he said.

Global demographic trends also will reshape the transportation industry. Today, half of the world's population lives in cities, he noted. That will grow to more 60 percent by 2030. Also, 80 percent of wealth by that time will be in cities. "So what types of transportation do we need for these cities? It will be very different from what we have here in the Midwest," Dr. Smyth said. "So again, it will be a portfolio of solutions."

Over the past five years, GM has been developing such a portfolio, Dr. Smyth said. GM is launching several electric vehicles in 2012. The Volt, which is between a plug-in hybrid and a pure electric vehicle, was to be rolled out in seven U.S. states in the fall of 2010 and launched nationally over the following 12 to 18 months. GM also is working on fuel cell and hydrogen vehicles. "When you look at the portfolio, we already have all the options," he said. "It is not about having niche plays.

The big question regarding energy sustainability and CO2 it is not about niche plays, he said. "It is about how we transform the vehicle fleet. Car companies such as GM "already have developed the niche plays," he said.

"Range anxiety," however, is a real issue with consumers. Dr. Smyth said GM learned from its experience with the EV1 electric car program in the 1990s, when range anxiety was a huge issue. "This is an area where we are not compromising the utility of the vehicle for the customers," he said.

Lithium-ion batteries also are a challenge. Dr. Smyth noted that a current battery pack producing 16 kilowatt hours of power with a 40-mile range weighs 400 pounds, is six feet tall and six feet six inches long. "That is a lot of mass and a lot of volume that is required even for that range," he said. GM offers a 100,000-mile warranty on the battery pack.

Regarding charging, Dr. Smyth said GM's view is similar to that of Dr. Sperling. Most of the time, electric vehicles are at home. "Having the right infrastructure at home, especially at 240 volts, makes the most sense," he said. The next priority is charging at work. He agreed that public charging "is of very limited value."

Electricity is a low-cost source of energy for a vehicle, Dr. Smyth noted. At $2 per gallon, gasoline costs 7 cents per mile. At $4 per gallon, it costs 13 cents per mile. If one pays 11 cents per kilowatt hour for electricity, by contrast, the cost is 3 cents per mile.

Another consideration is cost of ownership, Dr. Smyth said. Assume an average hybrid offers fuel savings of 30 percent to 40 percent. "What does that actually mean to the rational customer?" he asked. "That is the challenge we have today." A typical mid-sized family car that gets 35 miles per gallon and is driven 12,000 miles consumes about 400 gallons of gasoline a year, he said. If gas costs $3 a gallon, a hybrid saves around $300 in fuel. "Even at $6 a gallon, that is not a lot," Dr. Smyth said.

It is very hard to make an economic case to consumers by focusing only on fuel economy, he said, especially when one factors in the $3,000 to $6,000 additional cost of buying a hybrid rather than a conventional car. Dr. Smyth said he agrees with Dr. Sperling that it is necessary to maximize all of the other benefits of electric vehicles. "When you look at the actual cost savings, they are really quite limited," he said. "And that is a very big challenge, especially as you go through that Valley of Death."

GM has done quite a bit of recent research looking not only at the first few generations of electric vehicles, but also at Generation Three vehicles expected to arrive around 2020 and 2025. "You are still looking at very costly options" versus internal combustion engines, Dr. Smyth said.

GM view the Volt as an important learning experience, Dr. Smyth said. GM engineers have put the Volt through around 1 million miles of tests so far. Battery packs have been tested for more than 4 million hours. These tests have included driving the cars through water troughs. The Volts and batteries "have passed with flying colors," he said. "Clearly, this is a full-utility, no-compromise vehicle."

GM is working on a full range of key components for electrification. In March 2010, it opened an extensive battery laboratory in Warren and is planning an $8 million expansion, Dr. Smyth said. GM also is working on electric motors and power controls and will be exploring fuel cells, he said.

The company collaborates with universities on R&D. "A lot of research is still required," he said. GM must focus on performance, real-world drive cycles, and battery life, and other areas, he said. "While we have been working on internal combustion engines for 100 years, we are still in our infancy with regard to electrification," he said.

The $2 billion in federal stimulus funding "has done a lot to accelerate our capability in North America," Dr. Smyth said. GM invested $43 million in a manufacturing plant in Michigan's Brownstown Township, with the government providing $106 million, he noted. GM also invested $246 million in motor and electric-drive facilities. It received $105 million in federal funds for a plant in White Marsh, Md., to build high volumes of electric motors starting in 2013, he said.

GM is working on infrastructure needed "to make this comfortable for customer," Dr. Smyth explained. The company has been working with around 300 of North America's 3,000 utility companies, many of whom have different standards, he said. It is rolling out charging facilities at seven U.S. sites in 2010 and will go national in 2011. The industry must continue to work with utilities and the government to understand how to connect to the national grid, he said.

Collaboration will be crucial. The industry has gone through the initial learning stage of R&D. "The second learning stage is commercialization. It is the Valley of Death, and it won't be a narrow valley," Dr. Smyth said. "We have to work with the government and the utility companies to make sure we are producing the right technologies for our customer. We need the supply chain. We have to make it affordable to the customer."

Collaboration with universities also will "not just the next generation of engineers, but the current generation of engineers," Dr. Smyth said. "Again, it's not about niche products. We have them. It is about transforming the whole vehicle fleet and how do you do it quickly."

EARLY ADOPTION OF HYBRID VEHICLES

Bill Van Amburg
CALSTART

The truck industry may be a niche market for electric vehicles. But it offers a good case study of "how we might get things moving" in the U.S., said Mr. Van Amberg, CALSTART's senior vice president. It shows how advanced technology can be rolled out by focusing on a market segment in the commercial space, rather than the broad market. The important thing to building a market is to "get advanced vehicles out there," Mr. Van Amburg said. "You have to spur demand by supplying product."

The experience of electric trucks also illustrates "the power of public-private partnerships in filling the gaps," he said. "We are really good at R&D, but we drop the ball when it comes to getting things into pre-production." The model used by CALSTART to promote advanced technologies for commercial vehicles may offer lessons for other market segments, he said.

CALSTART was founded in 1992, during the last big recession, Mr. Van Amberg explained. California was facing a big brain drain, particularly in aerospace, defense, and other high tech fields. "We looked at what Michigan is really looking at: how to hold onto highly skilled talent in key industry segments and transition them into other fields." California focused on clean, advanced transportation.

Today, CALSTART has 130 corporate members. One-quarter of members are in the upper Midwest, Mr. Van Amburg said, because that is the "manufacturing corridor" of the U.S. and is "where the transition is happening." The "meat and potatoes" of CALSTART's membership are small and midsized companies because that group includes most of the technology innovators, he said.

Many policy makers are unaware that advanced technologies are moving into the commercial truck market, "an area not thought of as advanced technology," Mr. Van Amburg said. Hybrids now account for about 40 percent of the new market for transit busses, he said. "We also are seeing real movement, a real transition, in the truck world to advanced technologies," he said.

Mr. Van Amburg displayed photos of an array of small and light trucks, from Coca-Cola and FedEx delivery trucks to semi-trailer trucks. "Every photo is of a hybrid," he said. "What looks like a pretty conventional truck, is actually is a hybrid electric, hybrid hydraulic, all-battery electric, or a plug-in electric truck." Seventy percent of these vehicles are made in early, low-volume manufacturing, he said. "They aren't prototypes. They are actually moving into full-scale manufacturing."

This has "really been a sea change" in the truck industry, Mr. Van Amburg said. In many cases, he said, it is being led by the customer, who sees the value in getting better fuel economy.

Tougher regulations also are driving truck electrification. On May 21, 2010, President Obama announced there will be new carbon and fuel-economy standards for trucks[29] that should hit by 2014, Mr. Van Amburg noted. These are the first such rules for commercial vehicles.

There also are new ozone rules "that are intriguing," he said. The EPA has announced it will strengthen current ozone rules to meet health standards. The

[29] See White House Office of the Secretary, "Presidential Memorandum Regarding Fuel Efficiency Standards," May 21, 2010 (<*http://www.whitehouse.gov/the-press-office/presidential-memorandum-regarding-fuel-efficiency-standards*>).

rules would drop ozone limits to no more than 0.06 to 0.07 parts of per million over eight hours of pollutants such as nitrous oxide and will be phased in over 20 years. The current limit is 0.075 parts per million. The EPA will issue final standards Aug. 31, 2010, and each state must outline plans to meet them by 2013.

CALSTART estimates that hundreds of U.S. counties—perhaps as many as 650--will likely fall into non-compliance due to the tougher EPA standards. "This is a little under the radar at the moment," Mr. Amburg said. "Certain areas are really going to get hammered." Southern California is an example. "The district in Southern California is now looking at this and saying it could be so bad that they may have to cease fossil fuel combustion in the basin," he said. While that may be "scare tactics," officials are calling for a renewed push for zero-emission transportation in Los Angeles, Mr. Van Amburg said.

Southern California is starting to look at electrifying freight corridors. CALSTART just started a project with the Los Angeles transportation authority and ports to look at zero-emission freight movement through a 17-mile corridor of I-710 running from the area's major sea ports—the arrival points of 40 percent of cargo entering the U.S.—to the rail beds.

The Department of Defense also is pushing to improve fuel efficiency, Mr. Van Amburg noted. The DOD's Quadrenial Defense Review cites climate change as a global threat for the first time, and has set a target of cutting energy use at non-combat facilities by 34 percent by 2020. That translates into a 2 percent annual reduction in fuel use, he noted. Most of the impact will be on medium- and heavy-duty vehicles.

A shift toward high-efficiency trucks will create U.S. jobs, Mr. Van Amburg predicted. "We often think these high-technologies are only going into cars, but there but is a huge job and economic savings potential in the medium- and heavy-duty space as well because that is where all the freight movement is," he said. A new report by the Union of Concerned Scientists and CALSTART predicts 124,000 new jobs and $24 billion in savings to the U.S. economy if policies support high-efficiency trucks.[30] The biggest gains will come in the Midwest.

Hybrid trucks now on U.S. roads are found in all weight classes. Work on hybrids for long-haul trucks also is underway, he noted. Truck makers "are looking at how to solve some of the specific market niches that can use these technologies, not the whole marketplace," he said. Kenworth, Peterbilt, Navistar, and Freightliner all have regional haul hybrid tractors and are developing next-generation electric and hybrid versions.

[30]Don Anair and Jamie Hall, "Delivering Jobs: The Economic Cost and Benefits of Improving the Fuel Economy of Heavy-Duty Trucks," Union of Concerned Scientists and CALSTART, 2010. (http://www.ucsusa.org/assets/documents/clean_vehicles/The-Economic-Costs-and-Benfits-of-Improving-the-Fuel-Economy-of-Heavy-Duty-Vehicles.pdf).

FIGURE 11 All electric drive trucks emerging.
SOURCE: Bill Van Amburg, Presentation at July 26-27, 2010 National Academies Symposium on "Building the U.S. Battery Industry for Electric Drive Vehicles: Progress, Challenges, and Opportunities."

Some companies also are unveiling demonstration all-electric heavy-duty vehicles. In fact, Mr. Van Amburg said, "the first all-electric vehicle in the United States by a major manufacturer is coming from the truck industry." Navistar is testing a unique dual-mode hybrid design that uses an electric drive for speeds up to 48 miles per hour and a blended mode at higher speeds, Mr. Van Amburg explained. It is delivering its first vehicles to FedEx. The second all-electric will be from Freightliner, and the third probably will come from either GM or Nissan, he said. Vision Industries is demonstrating a fuel cell hybrid electric drayage truck.

Plug-in trucks also are arriving in specific niches. For example, trucks are being fitted with extra rechargeable battery packs. This meets a market need for trucks that drive to works sites and idle while tools are operated. The engine can be shut off and the tools powered by the battery. "A lot of people are looking at this in a hybrid configuration and with an energy-storage chassis," Mr. Van Amburg said. A similar concept is being used in Class 3 trucks,[31] which are popular with fleet operators. Although they are light-weight trucks, they consume nearly as much fuel as Class A heavy-haul trucks because there are so many of them, he noted.

---

[31] A Class 3 truck is a light truck with a gross vehicle weight rating of 10,001 to 14,000 pounds.

Hydraulic hybrids are another "intriguing new flavor," Mr. Van Amburg said. Some vehicles under development use an accumulator to store hydraulic energy. They will start appearing in late 2010, he said.

What is really needed to push the transition to electrified trucks "is a coordinated set of standards, policy incentives, and regulations across the whole continuum to the market," Mr. Van Amburg said. "We have generally done a good job at the first piece, the research and development. But we have kind of been dropping the ball moving into the next level of volumes in the market to not only launch products but also justify the investment by the manufacturers and suppliers."

For the past decade, TARDEC and CALSTART have explored applying advanced technologies in military vehicles through the Hybrid Truck Users Forum (HTUF), which was scheduled to hold its annual conference in Dearborn in September 2010. The task is to deploy the latest technologies "but without having to pay the Mil-spec,[32] one-off price," Mr. Van Amburg explained. A big question is "how to get the robust manufacturing industry that has the capabilities of meeting our demand." The partnership looks at dual-use opportunities for both the military and commercial vehicle markets, he said.

HTUF has focused on the end-user, Mr. Van Amburg said. "In the past what happened is that a lot of the technologies coming into the marketplace were pushed on the end-user rather than pulled by the end-user," he explained. So HTUF studied applications that could really add value to buyers. It also interviewed users in early "beachhead" market segments. Working groups began developing performance standards for military refuse, utility, parcel, regional heavy-haul, and other trucks. New working groups are addressing military non-tactical base vehicles.

One next big goal is "to get the military and government to actually start buying the vehicles and help create that early market," Mr. Van Amburg said. "There are an awful lot of vehicles in the government fleet. How do we get them to use this advanced technology?" The military working group hopes to deploy hybrid vehicles at bases within a year.

The HTUF project began when no major truck makers were involved in hybrids and there was no demand for fleets, Mr. Van Amburg said. Today, more than 2,000 such hybrid trucks on the road. Another 2,000 probably will be added in 2010, he said, and the market appears set to double every year for the next few years. "We've done it through these forums and working groups, targeting key application segments, and finding out what the user really is willing to pay for," he said. "What is the business case they are willing to accept?"

The team found "tremendous interest" in hybrid trucks "as long as vehicles are as reliable and useful as vehicles they are replacing," Mr. Van Amburg said. "We think we have really sped up the introduction into the truck world by two to

[32] "Mil-spec" is short for "military standard." The U.S. Department of Defense Standardization sets standards to achieve interoperability of equipment and meet certain requirements.

five years, depending on who you talk to." Hybrids are popular because they burn much less fuel than other vehicles, he said. "It's a big bang-for-the-buck investment area," he said.

In the initial phase, HTUF worked on building pre-production volumes and reaching agreements to purchase common vehicles based on common performance specifications. That information was fed throughout the military in order to produce "imbedded capabilities in the future hybrid," Mr. Van Amberg said. One interesting result of the dual-use approach is that most commercial hybrid trucks now have "silent watch" and "silent mobility" capabilities required by the military, he noted.

Now HTUF wants to build a sustainable market for hybrid trucks. "One thing we learned in the continuum to the market was to start with the first gap, to get people into pre-production volumes," Mr. Van Amburg said. "Let's get out of this one-off or two-off type of marketplace. Let's get into 20s, 50s, hundreds, and up to 500." By working with the truck industry, CALSTART learned that customers need help once they are into early production, he said, "because the prices are still high at that point."

HUFT has formed a new research center focusing on the next round of innovation, Mr. Van Amburg said. These areas include better energy storage, more efficient components, electric steering and braking for trucks, optimized engines, and integration with alternative fuels.

While the hybrid truck market is a niche compared to autos, the potential is impressive. Mr. Van Amburg estimated 30 percent of the world truck market can be "very addressable" for hybrid technologies by 2020. About 5 percent of regional heavy-haul tractors also can be hybrids. The projected numbers "aren't anything to stand up and scream about," but are sizeable nonetheless. More batteries are used in an electric truck than in a car, producing 50 to 100 kilowatt hours per vehicle.

Hybrids are at a "tipping point," Mr. Van Amburg said. "It is a very exciting time. They are right on the cusp of success in the marketplace," Mr. Van Amburg said. Some drop in the cost of energy storage and components can justify premium prices of hybrid trucks by savings in fuel and brakes in four to five years, he said. "But there is a chasm to be gotten across here, and we are working with policymakers to cross that chasm."

California has perhaps the nation's best and most innovative incentive program to promote electrification of trucks, Mr. Van Amburg said. It is managed by CALSTART. Rather than giving a tax credit, California pays half the incremental cost of buying a hybrid compared to a conventional truck. The subsidy is based on feedback from fleet-owners on what it would take for hybrids to make economic sense, he said. The state provides $10,000 per vehicle weighing from 10,001 pounds to 14,000 pounds and up to $35,000 for trucks weighing more than 33,000 pounds.

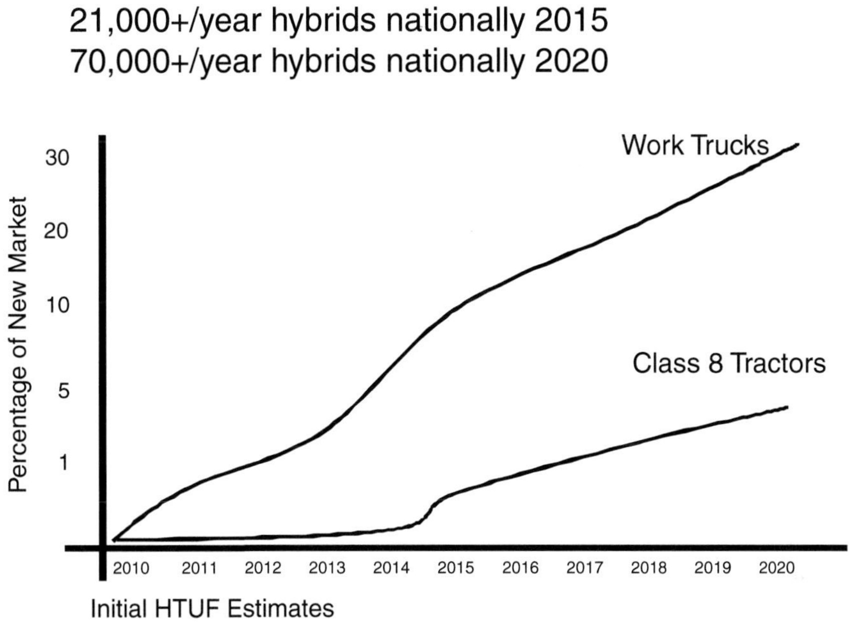

FIGURE 12 Introduction/impact framework.
SOURCE: Bill Van Amburg, Presentation at July 26-27, 2010 National Academies Symposium on "Building the U.S. Battery Industry for Electric Drive Vehicles: Progress, Challenges, and Opportunities."

The $20 million in the program is nearly used up, Mr. Van Amburg said. Some 600 trucks were purchased through the program, increasing the number of hybrids on the road by about 30 percent, he said. "These are the kinds of models that can actually work," he said. "We really have moved the needle in terms of advancing that technology."

CALSTART is working with U.S. Sen. Carl Levin on a federal incentive plan for hybrid trucks. Tax credits would apply when a truck is purchased in order "to reduce the capital cost directly to the commercial fleet," he said. Although tax credits don't work too well in the commercial market, Mr. Amburg said, "we're working with the tools that we have."

Mr. Van Amburg said he also has been working with a cooperative effort with manufacturers, academia, and government facilitated by TARDEC called the Advanced Vehicle and Power Initiative. The group is gathering ideas on "what would move the ball forward in saving fuel in the military and also create an industry and jobs," he said.

The initiative could be "greatly beneficial to the truck world and be really helpful to light-duty manufacturing," Mr. Van Amburg said. The AVPI calls for replacing 8 percent of the federal truck fleet each year with hybrids. "If one looks at all government truck fleets and an annual truck turnover rate of 8 percent, this would provide significant assistance to the early market and is something we would like to see more people talking about," Mr. Van Amburg said.

In sum, CALSTART's experience with trucks shows that "listening to the customer and focusing on how we target the key applications that will get the beachhead launched" was successful, Mr. Van Amburg said. "But over the long haul, we really need a coordinated plan that synchronizes our investment strategy, our incentives, and our requirements so that we can move out in a unified, long-term way." Besides long-term R&D, he said, "we need to target those gaps of moving into pre-production volumes faster and bringing in the user to target these key applications."

DISCUSSION

Dr. Sastry of Sakti3 remarked that it was good to hear discussion of the different sizes of battery packs needed, with some speakers talking about 50 kilowatt hours and others about 2 kilowatt hours. "The reality is that these will move up and down the scale," she said. Dr. Sastry asked what kind of infrastructure companies like GM and Ford have "to regularize, test, and control packs."

Some companies have invested in that capability, "but clearly there is lots of opportunity for additional test and development capacity for the electric battery, power machine, and power electronics, Mr. Kruse said.

GM has invested significantly in this area, Dr. Smyth said. "But we are still in the infancy. We really don't have real-world data yet." Issues such as high-temperature durability, low-temperature performance, how batteries fail, and accelerated testing are poorly understood, he said. "We have made a lot of progress, but a lot more needs to be done."

Mr. Kruse asked his fellow panelists how important they think petroleum prices are to electric-vehicle adoption rates.

Dr. Sperling said he doesn't believe fuel prices will stay really high or really low on a sustained basis. "This is probably about where they will be for a long time, except for spikes," he said. "In the end, it is more of a consumer perception thing. Yes, a high fuel price has a huge impact on peoples' psychology. At least in the price range we are talking about, it doesn't affect the economics much."

Mr. Van Amburg said gas prices may not fundamentally change the economic considerations of buying a hybrid car, but they make a big difference in commercial vehicles. "When you run the numbers on hybrids just at current costs without the platform costs going down, you can make the case for a three-year turnaround at $4 and $5 easily," he said. "The two biggest variables in commercial vehicles are the up-front costs of the vehicle and the fuel price. With these guys, it's their work tool. They know what they use in fuel, while most consumers really don't."

The case for electric vehicles can't be all about fuel, Dr. Smyth said. "If the price of fuel were high, certainly that would make it easy for us," he said. "But we have to go through this transformation, and we don't have a choice." We're pulling 85 million barrels of oil out of the ground every day." Oil in 2030 is projected to cost $105 or $110 a barrel, "and that is even with the current recession," he said.

The U.S. must diversify from petroleum, Dr. Smyth said. Currently, it is more 95 percent dependent on oil. "We must come up with low-carbon alternatives." Cellulose ethanol offers potential carbon savings of around 50 percent over petroleum with advanced internal combustion engines. "There aren't any 80 percent reductions out there," he said. "We really have to look at developing alternatives and developing them rapidly if we want to drive down CO2. We have to stay on this path, and that is why we have to be working with the government to make sure this is a success."

Dr. Good asked if anyone has made a serious attempt to model the entire system. "Let's assume you can get 30 percent electric vehicles in 20 years," she said. "Does anyone have a model of what the grid would have to look like?

Mr. Van Amburg said he has seen good models developed by utilities. "If you are shifting your charging to night-time, and you are using off-peak energy during that trough, it doesn't have that much impact on the overall production of the grid," he said. "The biggest issue is deploying individual vehicles on the grid because of the potential draw you are getting if you had a whole bunch of electric vehicles in one neighborhood. But based on the models I've seen, it wouldn't necessary make us get into a lot more production of electricity," he said.

Electric vehicle penetration of 5 percent is a very different problem than 25 percent, Dr. Good noted. "I don't think people have really run the right models yet. Even if I run off peak, I have to use more fuel to run the electricity. If I look around the country today, there are areas that have some extra capacity. But there isn't that much extra capacity running loose. I don't know that we have actually made a good model, and it seems we ought to do that fairly soon so we have answers to people about how we are going to provide that electricity. If you are trying to rev this up to 25 percent in the next 10 years, you had better get on that problem now."

It will be a very long time before 25 percent of cars are electric, simply because of the rate at which vehicles in use turn over, Dr. Sperling replied. "That is a long, long way off, and we do have models on that."

A more critical issue is the effect on local transformers, he said. "You have that neighborhood clustering effect that can be very disruptive," Dr. Sperling said.

Panel V

Building the Battery Workforce

Moderator:
Bill Harris
Science Foundation Arizona

The Michigan Economic Development Corp. deserves a lot of credit, said Mr. Harris, who prior to Science Foundation Arizona directed Science Foundation Ireland and served at the National Science Foundation for 18 years. "Looking at this from a distance, what you all are doing and what this state is doing is impressive," Mr. Harris said. "What we have to figure out as a country is how to seriously work together to build these capacities and build this future. Otherwise, we will struggle going forward."

Previous panels discussed political and industrial leadership, Mr. Harris noted. This panel addressed the workforce needed for an advanced battery industry. "One of the key aspects of the process is the 21st century workforce," he said. "And I think everyone here knows the country is struggling with a K-12 education system that is weak."

The U.S. used to lead the world in producing people with advanced degrees, Mr. Harris noted. It also "used to be thought of as the best engineering country in the world," he said. "Now we wonder across the country whether our kids can add and subtract."

This panel features a company that deals with workforce issues and a professor who deals with the production of students, Mr. Harris explained. "Hopefully, we can engage in a conversation about whether these things are matching up properly," he said.

The first speaker was Robert Kamischke, the chief financial officer and chief information officer of lithium-ion battery storage systems manufacturer EnerDel. One thing he liked about Mr. Kamischke's background, Mr. Harris said, is that he was an executive of General Motors' EV1 electric car program in the 1990s. "Years ago, I had the privilege of driving that little car, liking it a lot, reading the book about it, and seeing it towed away," Mr. Harris said. He said he hoped Mr. Kamishcke will be able to comment on whether the U.S. electric-vehicle program is now on the right track and about the workforce.

The next speaker, Simon Ng, "has a whole list of titles," Mr. Harris said, "but I think the one he probably likes best is distinguished faculty fellow and professor of chemical engineering and material science." At Wayne State University, Mr. Ng also is director of alternative energy technology, director of a national bio-fuels energy laboratory, and interim associate dean for research.

WORKFORCE NEEDS AND OPPORTUNITIES

Robert Kamischke
EnerDel

Mr. Kamischke noted that this is his "second time around" in the movement to electrify the transportation sector. "When the opportunity came to join again, I asked myself why I would do that," he said. "I felt it was compelling that we should work toward energy independence for our nation. Secondly, I was very interested in helping create a sustainable manufacturing base for our country."

EnerDel is more than just an advanced battery manufacturer, Mr. Kamischke said. EnerDel is "a complete lithium-ion solutions provider," he explained. The company is a fully owned subsidiary of Ener1. Its other subsidiaries are EnerFuel, a developer of low-temperature proton exchange membrane (PEM) fuel cells, and NanoEner, which develops processes for applying active materials on electrodes.

The company's key partners include the Department of Defense, the Energy Systems Network based in Indianapolis, Argonne National Laboratory, Purdue University, Rose-Hulman Institute of Technology, the U.S. Advanced Battery Consortium, and Japan's Itochu. EnerDel supplies customers in consumer electronics, transportation, industry, power-generation, and the military, he explained. They include Nissan, Volvo, TARDEC, AC Transit, Think Automotive, and Portland General Electric.

EnerDel has built the first commercial high-volume cell fabrication plant in the United States, Mr. Kamischke said, thanks in part to $118.5 million in federal money through the Recovery Act. The plant was set to open in the fourth quarter of 2010. In November 2009, it was awarded a contract by TARDEC to work on the Humvee. It began commercial production of battery-pack systems for Think Automotive in May 2010.

Success in vehicle electrification will have important benefits for the U.S., Mr. Kamischke said. He noted that the transportation sector accounts for 70 percent of the U.S. trade deficit. Light-duty vehicles represent 40 percent of annual U.S. oil consumption, which is about 25 percent of the U.S. trade deficit. "The U.S. is not the only country that imports more oil than it produces," he pointed out. "China imports more than 52 percent of the oil it uses, and you can see the urgency with which they are moving toward electrification of their transportation sector to mitigate that." Oil price volatility is another problem for the U.S. If one compares oil prices to electricity prices, oil is seven times more volatile, Mr. Kamischke noted.

EnerDel is "very appreciative" of what the U.S. and European governments have done to support vehicle electrification, he said. The company is selling the first auto applications of its batteries in Europe to Think Automotive. That Swedish company soon plans to assemble its electric vehicles in the U.S.

The workforce needs of the advanced battery industry vary across the supply chain. In electrode manufacturing, skilled workers are needed for the mixing, coating, calendaring, and slip-punch processes, Mr. Kamischke explained. To make cells, they are needed for the dry room, electrode-stacking, assembly, and formation processes. Skilled workers also are required for pack assembly and testing. The training and education requirements for manufacturing positions range "all the way from high-school degrees to Ph. Ds.," he said.

Engineers will require four-year degrees, while some working in areas such as advanced materials, chemicals, modeling, and simulation will need advanced degrees, he said. The company also will need engineers to design electrical circuits, mechanical systems, and software to run the systems. The U.S. skills gap mainly is with advanced-material and chemical engineers, he said.

Most of EnerDel's workforce, however, will be in "middle-skill" operations work, Mr. Kamischke said. The ratio of middle-skill to high-skill workers, in fact, is around five-to-one. In some of the more demanding manufacturing processes, such as with cell and electrode fabrication, EnerDel will look for workers with two-year applied sciences degrees. "They will be focused on the industrial technology path, advanced manufacturing, or engineering technology," he said.

Demand for "middle skill" workers may exceed supply. Currently, 56 percent of demand for workers in Indiana falls in this category. This ratio is likely to remain stable through 2016, Mr. Kamischke said. However, only 45 percent of Indiana's workforce qualifies as "middle skill."[33] "We have demand for those jobs outstripping the workers," he said. "This is one area of concentration EnerDel is working to solve with the state of Indiana and our educational institutes."

Compared to the rest of the nation, Indiana has a relatively low percentage—less than 40 percent--of adults aged 25 to 64 who have at least an associate's degree, Mr. Kamischke said. Indiana ranks ahead of only West Virginia, Louisiana, Nevada, and Tennessee. [34]

An innovative Indiana community college called Ivy Tech is working to resolve this gap. Ivy Tech has 23 campuses across the state and 130,000 students. "As we see it, Ivy Tech will be part of the backbone of building this emerging middle work force for the renewables age," Mr. Kamischke said. "They are proactively teaming up with industry to create the workforce to meet this new demand." The school offers an associate's degree in applied science, for example, with focuses on industrial technology, advanced manufacturing, and engineering technology.

---

[33] Data from Indiana Department of Workforce Development and U.S. Census Bureau.

[34] Data from Indiana Cluster Skills Labor Market Monster Study.

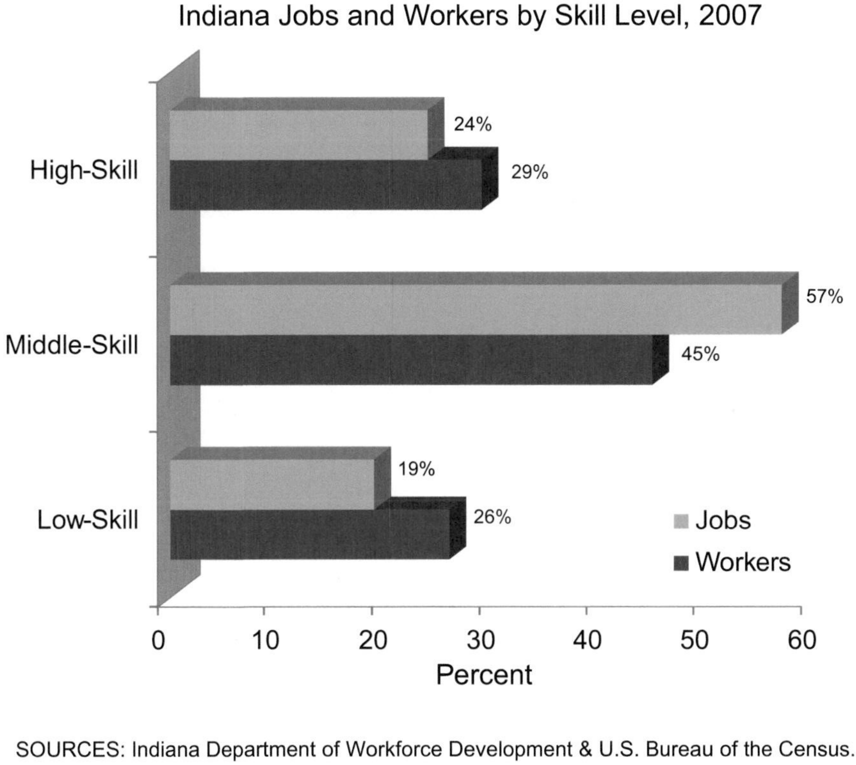

FIGURE 13 Supply and demand for middle skill jobs.
SOURCE: Robert Kamischke, Presentation at July 26-27, 2010 National Academies Symposium on "Building the U.S. Battery Industry for Electric Drive Vehicles: Progress, Challenges, and Opportunities."

Ivy Tech also is offers new degree concentrations for emerging industries. Indiana has one of the largest wind farms east of the Mississippi, for example, located off I-65 near Chicago, Mr. Kamischke noted. So Ivy Tech and the state are setting up degree programs in electric-line construction, power-plant technology, natural-gas technology, utility-scale wind turbines, and home-integration technology for auditing residences and tailoring solutions to save energy and reduce cost.

For the transportation sector, Ivy Tech is establishing curricula for electric vehicles, recycling, and first-responder training. A DOE grant helped fund this effort, he said. Ivy Tech also is working to develop a nationally recognized certificate for electric transportation technicians. There also will be a need for specialists in reusing and recycling batteries for other uses. Mr. Kamischke noted that there will be many more batteries than vehicles on the road as electrification expands. "There is going to be a need to refurbish and rehab that battery pack for another vehicle or grid application," he said.

Such training efforts are needed to keep America competitive, Mr. Kamischke said. "As other nations explore green-technology solutions, America must focus on broadening our education opportunities for young adults in science and manufacturing technology," he said. Continued investment in advanced transportation technologies will create "a rare surplus of jobs in many regions," Mr. Kamischke said. Therefore, state and federal governments should focus on middle-skill jobs.

TECHNICAL TRAINING AND WORKFORCE DEVELOPMENT

Simon Ng
Wayne State University

Dr. Ng opened the session by stating that Michigan's need to train a highly skilled workforce is obvious. When electric-drive vehicles reach 5 percent to 10 percent of the market, "just imagine the number of skilled workers we will need to fill those jobs," he said. He also stated a need to re-train other workers in the auto industry so they will understand how to transition from an internal combustion engine to a battery and electric-motor system In addition to automakers, many battery manufacturers are coming to Michigan and they too will need many skilled workers to produce millions of battery packs.

Wayne State is developing a comprehensive set of degree programs aimed at filling anticipated needs for electric-drive technology and batteries. The collaboration with Macomb County Community College and NextEnergy[35] is one of three funded by the DOE in Michigan. Guidance for this program is critical, and so the advisory board for Wayne State's program includes executives from GM, Ford Motor, DTE, TARDEC, AVL, and Compact Power, among others.

The program's mission is very straightforward. "We want to design a program so that we can prepare our current and future workforce with the education and skills necessary for the advancement and maintenance of electric-drive vehicles," stated Dr. Ng. Wayne State dubs its program E3, which stands for electrification, economy, and education; it is about electrifying the economy and educating the workforce

Wayne State's offerings include a master's degree in electric-drive vehicle engineering and a bachelor's in electric transportation, Dr. Ng explained. It also offers is an undergraduate concentration and graduate certificate in electric-

[35] Next Energy, launched in 2002, is a nonprofit organization based in Detroit dedicated to bringing promising technologies to maturity accelerating commercialization and scale production. It manages research projects in power-generation, transportation, and fuel technologies.

vehicle engineering. Many technician-level operators also are needed and Wayne State's partner, Macomb Community College, offers an associate degree program in automotive technology and electronic engineering technology.

To design the curriculum, Wayne State relied on input from the electric-vehicle industry and OEMs. It also studied best practices in electric vehicle-related curricula from around the world. Dr. Ng recently visited key Chinese universities and noted that China is putting a tremendous amount of investment in industry and universities. It became apparent that, in looking at their universities and infrastructure, China is determined to train and prepare their workforce for the electric-vehicle industry. Dr. Ng noted that the Obama administration recognized the importance of workforce development and invested in higher education in developing electric vehicle education programs. Without this investment, the U.S. might otherwise lose its leadership position in the critical electric vehicle industry.

After obtaining initial data, the next step was to design a cohesive set of graduate, undergraduate, and technician-level courses. Making sure all courses contain interactive laboratory modules was another priority. "After all," Dr. Ng stated, "we are engineers. We learn by doing things."

The degree programs are being implemented with synchronous and asynchronous web-based distance-learning technology, with the anticipation that they will have a national impact on educational programs. The curriculum will be validated by industry, with the expectation that the degree programs will become accredited at some future date.

One objective of the program is that it be comprehensive. "As you all know," stated Dr. Ng, "the electric-drive itself is not just electrical engineering. It is not just mechanical or chemical. We wanted to pull all of these resources together so we involved faculty from electrical, chemical, mechanical, industrial, and alternative-energy technology to make sure we have a comprehensive approach to the curriculum."

A second objective is to be industry-oriented. The university hosted a workshop with a number of companies to learn what they need for their workforce. In addition, to make sure students have real-life laboratory experience, the program is working with companies to use their laboratories for practice and experiments.

Another objective is to have a broad impact. Therefore, it addresses every level, using a system known as "two plus two plus two." That means a two-year associate's degree, two years of engineering technology, and two years of master's level curriculum, Dr. Ng explained.

Eventually, the program will be translated for distance learning. The school is considering several strategies so that distant-learning students can also gain lab experience. One strategy is to develop simulations that can run experiments. Remote control is another approach. "Online students can be in Texas," he said. "But they can remotely control the instrumentation and actually run the experiment and analyze results." Wayne State also may require distant students to come to campus one week per semester to focus on experimentation.

The master's in electric-drive vehicle engineering program consists of 32 credit hours. There are thesis and non-thesis tracks. Some workers, however, may not have the time or energy to pursue a complete master's degree. For them, Wayne States offers a graduate certificate program. The certificate requires 12 credit hours, which is equivalent to three courses. "Essentially, they can work and take courses at night so that after a year or so they can receive a graduate certificate," he explained. The bachelor's program has 64 credit hours for the third and fourth years.

The electric vehicle-engineering program was launched at the beginning of 2010. It formed an advisory board and launched a Web site in April 2010. All of the degree programs have now been approved by the university, so that the program can begin courses in the fall of 2010, Dr. Ng said.

The E3 workshop conducted by Wayne State drew about 120 attendees, with representatives from 12 universities and community colleges and some 30 companies, Dr. Ng noted. The workshop had three tracks—batteries, vehicle integration, and infrastructure. One clear message from industry for the battery curriculum was that it is necessary to have a fundamental course on electrochemistry. "The way they look at it, you really have to have a system engineering approach, but get down to the molecular level," Dr. Ng said. "Then there is the cell level, the pack level, then the system level." Battery recycling and manufacturing were other major needs cited by industry representatives.

A number of new laboratories are being developed for the electric-vehicle program. The energy-storage laboratory is separated into three levels, Dr. Ng explained. Cell fabrication is more at the molecular level. Students learn how to make new materials, as well as how to make cathodes and anodes and how to put a cell together. The next level is to learn to characterize cells and learn different techniques to study subjects like thermal management. Students also study characterization of battery packs and modules in order to prepare them for potentially working for Ford, General Motors, and Chrysler, having considerable experience in testing battery packs.

A second laboratory is dedicated to electric propulsion, where various vehicle drive cycles can be studied. Braking, acceleration, and electronic controls can be simulated for different kinds of hardware and systems. The lab will enable students to engage in hands-on learning and to understand different issues of integrating the vehicles. A third lab allows for experimentation with electronic controls and studying interactions between batteries and electric motors.

The fall 2010 curriculum includes courses on the fundamentals of electric-drive vehicles and battery and battery systems. In infrastructure, there are courses on power electronics and vehicle-charging infrastructure. There are modeling courses for electric vehicles and power trains, as well as courses on design, production, and infrastructure development. A course on energy economics and policy will provide students with a comprehensive understanding of the impact of energy and policy on the development of electric vehicles. An advanced course on material sciences for batteries is being taught by an expert from General Motors. There also are advanced topics in electric vehicle-control optimization, embedded systems, thermal management, and hydrogen production and storage.

Day Two

Welcome and Introduction

Andy Levin
Michigan Department of Energy,
Labor and Economic Growth

Michigan has taken a lot of effort to understand the workforce and training needs of the electric vehicle sector, said Mr. Levin, acting director of the state's Department of Energy Labor, and Economic Growth. The agency "is sort of like an Energy Department, Labor Department, and Commerce Department all combined in one," he explained. "While that is a lot to keep track of, there are tremendous advantages in having the workforce capacity and energy capacity of the state working together in one department."

Michigan's effort is exemplified by the No Worker Left Behind Initiative, which Mr. Levin described as "the most aggressive workforce training program in any state of the nation." Since the initiative began in August 2007, more than 135,000 Michigan workers have undergone for in-demand degrees and certificates, he said.

One need not meet the qualifications of federal unemployment insurance to qualify. If a person is unemployed, about to be unemployed, or even is working but has a family income of $40,000 a year or less, he or she is eligible for up to $5,000 a year or $10,000 for two years of free tuition and other support at any Michigan community college, university, or other approved training program. "This program has been so popular, and such a huge success, that we will have 60,000 people in training in the workforce program that started July 1 [2010]," Mr. Levin said.

Rather than training new workers, the focus is on "really up-scaling our workforce," Mr. Levin explained. As many as three out of four participants are in training programs of one year or longer. Many are earning associate or bachelor's degrees or using money to finish master's degrees, he said.

Michigan conducted serious research on its "green economy" and "green workforce," Mr. Levin said. "There is so much hype and so much fluff and so much vagueness about the green workforce," he said. He noted that the department recently won a national award for "the best piece of labor-market demographic information research."

The department identified and defined five green sectors, one of which is "clean transportation and fuels." More than 6,000 Michigan employers returned surveys. The study found Michigan has 109,000 "real" green private-sector jobs, Mr. Levin said. Of them, 97,000 are direct jobs, such as welders making components for a wind turbine, he said. Twelve thousand are "indirect" jobs, such as janitors, accountants, or general counsel staff whose jobs wouldn't exist without the green-production work. Green jobs accounted for 3 percent of private-sector employment in Michigan, he said.

Employment in the green sector is growing fast. Between 2005 and 2008, overall employment in Michigan's private sector shrank by 5.4 percent, Mr. Levin said. Green employment grew by 7.8 percent, adding 2,200 new jobs, 700 of them in companies that did not exist in 2005.

A recent study found that 90.6 percent of the state's clean transportation and fuels jobs are in southeastern Michigan, with a small but growing workforce in the southwestern part of the state. What's more, 55 percent of green jobs in southeast Michigan are in clean transportation and fuels. This high concentration "is something you will not find in any other region of the United States," Mr. Levin said. "That is how important this sector is to our economy."

To train workers, Michigan launched a $6 million green jobs initiative in 2008 to work with employers, identify sectors, and create Michigan "skills alliances" in particular areas where employers want to train their workers, Mr. Levin said. One example is the Michigan Emerging Market Skills Alliance. It works with small tool-and-die companies and suppliers that must diversify. "Typically, they are one-trick ponies that supplied one company," he explained. "And now they clearly need to diversify, often into batteries, wind, solar, and things like that."

Another such program is the Michigan Academy for Green Mobility. It trains engineers for vehicle electrification, Mr. Levin said. "This is typical of what we do," he said. "If employers don't know about these things and are interested, talk to me. We want you involved in this."

Sometimes public-private training programs are launched at the initiative of one company, he said. Executives of Ricardo Engineering met at the governor's office around fours ago, for example. "They said, 'We are going to need hundreds and hundreds of engineers who know how to work on hybrid and electric vehicles, and we ain't got them. Not just Ricardo. The auto industry,'" Mr. Levin recalled.

State officials convened representatives of GM, Ford, and Chrysler, as well as Japanese companies and university officials. "We all struggled together about whether this is right and what we will do," Mr. Levin said. The state asked employers to identify precisely what kind of training was required. "We are talking about engineers who already have bachelors or master's degrees, or maybe even a Ph. Ds," he said. Whether they had lost their jobs or were still working, they needed new skills to work on electrification.

Wayne State University and Michigan Technological University won competitive bids to serve as lead universities to run training programs. Three hundred skilled workers have gone through the program so far, he said. Now the Department of Energy, Labor, and Economic Growth is talking to battery manufacturing companies setting up in the state about training workers, he said.

Michigan has aggressively sought funding for training efforts, Mr. Levin said. The state won a $5.8 million training grant for the energy sector. He praised the MEDC for "developing the strategy for putting Michigan at the center of this electrification push."

Mr. Levin said he wants companies in the electric vehicle industry "to know how much of the workforce capacity of the state is at the disposal of this effort." The state wants employers "to come to the table to work with each other and figure out whether there is a baseline of training we need for our workers in the same market," he said. "That is what we have been doing, and we want to keep doing with you all."

Mr. Levin said he is glad the National Academies chose Michigan for this symposium. "I also am glad to see how far we have come and to see that people are really pushing through these discussions about how to grow this industry quickly and how to make this a real driver of Michigan's economic resurgence," he said.

Panel VI-A

Federal and State Programs to Support the Battery Industry

Moderator:
Charles W. Wessner
The National Academies

THE DEPARTMENT OF ENERGY BATTERY R&D PROGRAM AND GOALS

David Howell
U.S. Department of Energy

A lot has happened in the Department of Energy recently in the energy-storage area, Mr. Howell said. "If you asked me to describe energy storage at the DOE two years ago, it pretty much would have focused on the vehicle-technology effort," he said. Since then, significant funds have been allocated through a number of DOE offices and programs.

The DoE's research for advanced vehicle batteries is run by the Office of Energy Efficiency and Renewable Energy Vehicle Technologies. The Office of Electricity already had been funding about $4 million in research for grid storage, Mr. Howell explained. President Obama has requested boosting that funding to $40 million in Fiscal Year 2011. The Advanced Research Projects Agency-Energy (ARPA-E), meanwhile, funds high-risk research seeking technology transformations in energy storage. The DoE's Office of Science "has kind of re-entered the game" through chemical research projects relating to batteries.

In 2009, the DoE's R&D budget for energy storage research amounted to around $100 million. Of that, Mr. Howell noted, $76 million went to vehicle technologies. Funding for energy-storage research will grow to $209.7 million under the FY 2011 funding request, with greater portions set aside for areas like the grid. The transportation research budget will get a $20 million increase.

The Recovery Act infused another $120 million into storage-technology research. That sum does not include $1.5 billion in Recovery Act grants for battery manufacturing and $585 million for demonstrations of electrified transportation and grid projects, he noted.

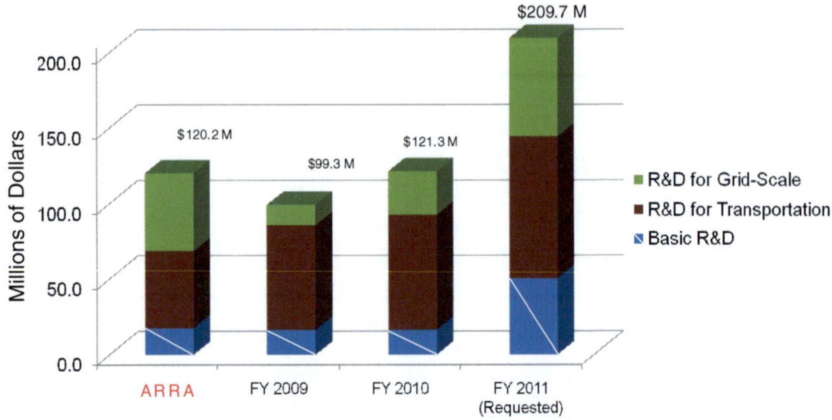

This chart does not include ARRA funding for advanced battery manufacturing ($1.5 B) or demonstrations ($400 M for transportation and $185 M for grid-scale)

FIGURE 14 Energy storage R&D funding from DOE and Recovery Act.
SOURCE: David Howell, Presentation at July 26-27, 2010 National Academies Symposium on "Building the U.S. Battery Industry for Electric Drive Vehicles: Progress, Challenges, and Opportunities."

Vehicle-technology research funded by the DOE spans the entire development chain. It includes R&D for advanced materials such as high-energy cathodes and high-voltage electrolytes, high-energy and high-power cells, full system development and testing, and commercialization.

Of the $76 million FY 2010 vehicle-technology R&D budget, $44 million goes to plug-in hybrid electric vehicles, with $15.8 million going to conventional hybrids. The remainder is for "exploratory" research, Mr. Howell explained. Roughly half of funds go to national laboratories and universities for next-generation research. The rest goes to industry, he said.

The DOE has a "documented track record of success" in battery research going back to nickel metal hydride technologies, Mr. Howell said. The agency has worked with lithium-ion development since the late 1990s. Many recent lithium-ion commercialization successes have been supported by the DoE's U.S. Advanced Battery Consortium program, which includes lab research, diagnostics, and modeling of new materials to help understand failure rates and the electrochemical phenomena of different chemistries, he said. "It's a full-program effort," Mr. Howell said. "It culminates, hopefully, in commercialized technology."

Recent successes have included the high-power lithium-ion battery pack developed by Johnson Controls-Saft for BMW and Mercedes hybrids. Others are the lithium-ion cells that A123 is supplying to Hymotion's Prius conversion program and the battery pack Compact Power/LG Chem will supply to the GM Volt.

One reason behind this success is that the DOE develops detailed targets with its industry partners in the USABC, Mr. Howell said. "We don't just suck these things out of our thumb," he said. "We do a lot of analysis of how batteries operate in a vehicle. Then we come up with a set of performance goals that we manage our R&D projects toward."

An example of such targets is to develop batteries for hybrids with 25 kilowatt discharge pulse power in 2010. "Auto companies may want 22, 30, or perhaps 40, but the point is this gets us into space of how a battery is supposed to operate in these types of vehicles," Mr. Howell said. The goal is to push that power to 38 to 50 kilowatts by 2015 from plug-in hybrids and to 80 kilowatts for all-battery electric vehicles by 2020.

The DOE changed its focus in batteries to lithium in 2000 because nickel metal hydride technology was being commercialized, Mr. Howell explained. In particular, the DOE targeted lithium ion for conventional hybrids. "This was pretty much a success story for us," he said. Most of the lithium ion chemistries that are now mature have been demonstrated to work for 300,000 cycles over a 10-year life, he noted.

DOE Energy Storage Goals	HEV (2010)	PHEV (2015)	EV (2020)
Equivalent Electric Range (miles)	N/A	10-40	300
Discharge Pulse Power (kW)	25	38-50	80
Regen Pulse Power (10 seconds) (kW)	20	25-30	40
Recharge Rate (kW)	N/A	1.4-2.8	5-10
Cold Cranking Power @ -30 °C (2 seconds) (kW)	5	7	N/A
Available Energy (kWh)	0.3	3.5-11.6	30-40
Calendar Life (year)	15	10+	10
Cycle Life (cycles)	3,000	3,000-5,000, deep discharge	1,500 deep discharge
Maximum System Weight (kg)	40	60-120	300
Maximum System Volume (l)	32	40-80	133
Operating Temperature Range (°C)	-30 to +52	-30 to 52	-40 to 85

FIGURE 15 DOE and USABC battery performance targets.
SOURCE: David Howell, Presentation at July 26-27, 2010 National Academies Symposium on "Building the U.S. Battery Industry for Electric Drive Vehicles: Progress, Challenges, and Opportunities."

Cost reduction and improved tolerance are top priorities, Mr. Howell said. He said the record of USABC battery programs has been impressive. The cost of a 25 kilowatt hybrid battery pack has dropped by around two-thirds between 1997 and 2007, while densities and calendar lives have more than doubled. "We have increased energy density of these materials while at same time decreased their cost and increased their life," he said. "One point we often make is that if you are going to reduce the cost of batteries you have to increase their performance at the same time so that you don't divorce cost from performance."

Work on plug-in hybrids through the USABC began around two years ago. The first projects are nearing completion, Mr. Howell said. A123 has a contract to develop batteries using nano-phase iron-phosphate, for example. Johnson Controls-Saft, EnerDel, 3M, Celgard, and Entek also have projects underway. The USABC has selected 12 new proposals to develop advanced high-performance storage systems.

Progress also is being made on batteries for plug-in hybrids. Batteries for a plug-in hybrid with a 10-mile range now cost about $2,500. That should drop to $1,700 in 2012. Improvements in cycle time, battery life, weight, and volume also seem to be on track. "But when you go to higher mileage plug-ins and electric vehicles, the targets get a lot tougher," Mr. Howell said. "So we have to move beyond the next generation of lithium ion to meet the targets."

The Vehicle Technologies Program does a lot of modeling to understand the main cost drivers for batteries, he said. Three cost models are used. The USABC model is a detailed hardware model to calculate costs for different designs with validated cell performance. "Typically, when I give cost numbers I am using USABC cost numbers," he said.

Two others, the Argonne and TIAX cost models, assess what is technically feasible for a given chemistry with a given design. These models help researchers understand the impact of cell chemistry on cost, active material costs, and the effect of scale manufacturing.

Under current estimates, a plug-in hybrid battery pack produced at high volumes costs between $700 and $950 per kilowatt hour, Mr. Howell said. The Argonne and TIAX models project that reducing that cost to $300 per kilowatt hour is plausible, he said. Instead of being based on actual validated cell performance, Mr. Howell explained, these models estimate costs for certain technologies "if everything works right."

These analyses have produced other useful insights. While the costs of active cathode and anode materials are important, for example, they represent less than 15 percent of the total cost of a battery pack for plug-ins with 10- and 20-mile ranges, he said. As one goes up to higher driving ranges, however, the models show "we need to develop materials with higher specific energy," he said. More materials in the pack remain active, reducing the number of cells required, packing costs, and other factors.

Manufacturing scale also matters. Boosting output from 10,000 batteries per year to 100,000 cuts cost by 30 percent to 40 percent, Mr. Howell said.

Consumer cells, he noted, now cost as low as $200 per kilowatt hour. "At the cell level, very high volume helps you drive down the cost," he said.

The Vehicle Technology Program also works on materials and processing technologies through a program administered by Oak Ridge National Laboratories. Five contracts and cooperative agreements are underway with A123, Johnson Controls, Dow-Kokam, Planar, and Porous Power Technologies.

In 2009, the DOE began an effort to support suppliers of materials. Angstron Materials and 3M are developing advanced anode materials. Companies like TIAX, EnerDel, BASF, and A123 are working on safety improvements and high-volume, low-cost manufacturing techniques, Mr. Howell explained. The DOE and companies each contribute $17.8 million.

The Vehicles Technologies Program also funds extensive research into many areas of electrochemical cells, with $34 million a year going into 60 projects at 10 national laboratories and 12 universities. National labs include Lawrence Berkeley, Argonne, Sandia, and the National Renewable Energy Laboratory. Universities include MIT, the University of Texas at Austin, the University of Michigan, and the New York state universities at Binghamton and Stony Brook. Ten of those projects are devoted to analyzing and constructing cells. Twelve focus on electrolytes, 15 on advanced cathodes, and five on modeling.

These projects have produced a number of "success stories," Mr. Howell said. "One of the ways you can gauge success from your universities and labs is whether industry really cares," he said. Several technologies developed in the program have been licensed to industry, he noted. Toda and BASF licensed composite high-energy cathodes developed at Argonne, for example. Hydro Quebec licensed conductive electroactive polymers developed at the University of Texas, and EnerDel licensed nano-phase and lithium titanate chemistry developed at Argonne.

In future research, the program will concentrate on several key areas. One is high-capacity cathode materials, Mr. Howell said. A request for proposals will be issued in the fall of 2010. Other areas are new solvents and salts that allow for high-voltage electrolytes, high-capacity anode materials, and novel ideas to address problems with lithium metal.

Further funds from the Recovery Act enabled the DOE to beef up laboratory research. Some $20 million went into new facilities. Argonne, for example, got an $8.8 million grant to set up a facility to fabricate battery cell prototypes, scale-up materials production, and conduct post-test analyses. The Idaho National Laboratory, Sandia National Laboratories, and NREL set up testing facilities.

Another $100 million for energy-storage technology came from ARPA-E. Half of that sum went to "transformational" transportation technology research. Six ARPA-E energy storage projects were awarded in 2009. They went to Arizona State University, Envia Systems, FastCap Systems, Inorganic Specialists, Eagle Picher Technologies, and MIT. The Vehicle Technologies

Program and ARPA-E have "very well coordinated efforts" to figure out which space each agency should be in, Mr. Howell said.

For 2010, another 10 projects were selected for ARPA-E grants. They include projects in lithium-air batteries at the Missouri University of Science & Technology, an all-electron battery at Stanford, novel high-energy density lithium-ion cell designs and manufacturing modules at Applied Materials, and high-performance and ultra-low cost rechargeable batteries at MIT. "We're very excited about this type of activity," Mr. Howell said. "We do a lot of exploratory research, and we would touch on these areas. But this provides significant funds to increase the likelihood that we could pick up these kinds of technologies over two to three years and carry them into the vehicle technologies area."

Even if half of these research projects are successful, "that would be a big win for us," Mr. Howell said. Many of the projects related to technologies at the cell level that would yield more than 400 watt hours per kilogram of capacity. He said his programs tend to focus at the 300 watt-hour per kilogram level, "so it would be a great advance for us if some of these are successful."

Budgets also have been increased for the Basic Energy Sciences project, which focuses on fundamental materials research and explores electrochemical processes and concepts. "We're excited about this program too, because a lot of the fundamental knowledge feeding into our exploratory research program hopefully will be coming out of this activity," Mr. Howell said.

The Basic Energy Sciences Program provides $300,000 for three years to individual investigators or teams. It also funds 46 Energy Frontier Research Centers. Five of those centers "are directly applicable to the work we are doing in the battery R&D area for vehicle technologies," Mr. Howell noted. They include a project at Cornell to understand and control nano-structured interfaces for energy generation and storage and a center at Stony Brook to study fundamental chemical reactions in electrodes.

The DoE's Office of Electricity "is really ramping up their energy-storage effort," Mr. Howell said. Its budget for energy-storage for grid-scale projects shot up from $3.6 million in FY 2009 to $40 million in the requested FY 2011 budget. They include a $14 million project by Primus Power Corp., which is to deploy a wind farm in California's Central Valley and a $22 million wind-storage project by Duke Energy Business Services. The projects are of "particular interest" to the vehicle technologies program, he said.

Although the DoE's Vehicle Transportation Program "has a successful track record in developing technologies for commercialization," Mr. Howell said, he cautioned that "this has taken us a long time." The program has worked with lithium-ion for more than a decade. "It has taken a lot of effort to bring these technologies to the point where they are now commercially viable," he said.

Until recently, though, this program operated on a "shoestring budget" of around $25 million a year, he said. With the Recovery Act funds to establish the industry in the U.S. and the increased budgets of different DOE offices, Mr.

Howell said, "we hope it doesn't take us another decade to get the next generation of chemistries into commercialization."

DEPARTMENT OF DEFENSE BATTERY R&D PROGRAMS AND GOALS

Sonya Zanardelli
U.S. Army Tank Automotive Research, Development and Engineering Center

The U.S. Army is a demanding customer for advanced batteries. "Basically we want a battery that weighs nothing and is very small," said Ms. Zanardelli, the energy-storage team leader at TARDEC. "We also are looking at reducing logistical and fuel burdens because sometimes those costs can equal the costs of the batteries themselves or more." The Army wants longer calendar and cycle lives, Ms. Zanardelli explained. TARDEC is issuing new specification for combat vehicles that addresses battery management and performance and operation time.

The team's mission is to "develop and mature advanced energy storage technologies and transfer them to our vehicle platforms," she said. TARDEC has in-house testing and evaluation capabilities for various energy-storage technologies, she added, and is able to pre-qualify battery technology readiness levels.

TARDEC works with the Army Research Lab on proofs-of-concept and can take a technology all the way through to the systems level, Ms. Zanardelli said. It then gets technology ready to integrate into vehicles. TARDEC also provides "cradle to grave support" for all Army ground vehicles, she said. It supports all batteries used in the field and writes standardization specifications and maintenance procedures.

TARDEC collaborates with units across the Department of Defense and with other federal agencies. It works with the Communications-Electronics Research, Development, and Engineering Center (CERDEC), which handles all soldier applications. It also works with the Army Research Laboratory, which focuses on basic research, and research centers for the Navy, Air Force, and Marines. A "battery technology working group" comprised of experts in each service convenes four or five times each year "to look at programs across the board so that we don't have redundancy," Ms. Zanardelli said. Each service also has battery and material partners in industry.

The DOD and the DoE's Vehicle Technology Program are developing a memorandum of understanding on vehicle energy efficiency, she pointed out. The MOU will establish a cooperative relationship to identify, develop, and test energy-saving technologies. It will enable TARDEC "to leverage some DOE funding for our applications," she said.

The technology working group is developing a single rechargeable lithium-ion specification that will be applicable to all DOD components now being created, Ms. Zanardelli said. It will lead to standardization of tests and performance, reduce duplication, and cut the number of chemistries. This effort is to be completed in 2011.

The performance/power and energy requirements for Army vehicles are high. Ground combat vehicles require 45 kilowatts of quality, sustained electric power while they are stationary, Ms. Zanardelli said. Commercial hybrid vehicles require only 5 kilowatt hours and plug-in hybrids 16 kilowatt hours.

The combat vehicles must be able to conduct operations in silent mode while operating combat control systems, sensors, the integrated protection suite, and weapon fire controls for up to six hours or more with no transition delay. A battery-only solution, however, "is still very, very challenging to meet the high-power Silent Watch requirements given the practical battery state of the art and practical vehicle constraints," she said.

For the Air Force, energy-storage systems are generally restricted by their mass, Ms. Zanardelli said. Its major applications are for aircraft emergency power, small unmanned aerial vehicles (UAVs), and long-endurance UAVs. Small UAVs, she explained, run for eight to 10 hours and use hundreds of watts of power. Long-endurance UAVs must operate 40 to 50 hours and need thousands of watts of power. The Air Force is looking at high-energy fuel cells, high-power batteries, and high power-management converters, she said. "They are looking at this approach to reduce the weight, total life-cycle cost, and to enhance their overall capability," she said.

Naval vessels mainly have volume restrictions for batteries, Ms. Zanardelli said. The U.S. Navy is looking to use energy storage systems for unmanned underwater vehicles, shallow-water combat submersibles, submarine small distributed power systems, surface ship fuel economy, and surface ship pulsed and high power. Future surface ships will require 0.4 megawatts for active-denial weapons systems, 2 megawatts for laser weapons, and 30 megawatts for an electromagnetic rail gun planned for 2020, she said.

Soldiers also have special battery needs. Soldiers carry up to 100 pounds of gear and as much as 30 pounds of batteries to support mission-essential equipment, Ms. Zanardelli noted. "There is a real challenge to reduce the number of batteries a soldier carries," she said. CERDEC is looking at fuel-cell batteries and hybrid power sources.

TARDEC sees numerous applications for advanced batteries in ground vehicles. Major applications range from robotics and weapons systems to electromagnetic armor, in addition to applications in vehicles themselves. Most of TARDEC's advanced-battery resources are going into Silent Watch and starting, lighting, and ignition systems, Ms. Zanardelli said. For starting and lighting, "we basically need the battery to turn the engine on," she said, so most vehicle platforms require two to four batteries. Up to six or twelve are needed for Silent Watch.

Silent Watch is "one of the most stringent requirements for the Army," Ms. Zanardelli explained. Batteries supply the power for mission equipment when the main engine is off and the vehicle is stationary. The benefits of Silent Watch and Silent Mobility are that they emit a low thermal and acoustic signature and can provide a means to quickly generate power at peak electrical modes, better fuel economy, reserve power, and silent export power, she said.

TARDEC is looking at the business case for lithium ion in most vehicle platforms, Ms. Zanardelli said. It is looking at 12-volt drop-in replacement batteries compared to 28-volt replacements, for example. "How many pounds are shaved off a vehicle, and at what cost? What performance are we getting? Most of our programs focus on issues like that," she said.

The challenges with energy-storage technologies receiving the most attention by TARDEC include cell and system safety and reliability, Ms. Zanardelli said. It has developed a process the government and industry can use to make quantifiable measurements for identifying the technical maturity of battery chemistries, she said.

TARDEC has been taking many steps to reduce the cost of batteries through the ManTech program,[36] Ms. Zanardelli explained. One manufacturing technology project started in 2004 for Future Cat Systems focused on automating lithium-ion production processes and reducing the cost by half. Now TARDEC is transferring the knowhow to the Joint Light-Tactical Vehicle program, which is using lithium ion, she said.

The Integrated Target Acquisition System Program, which has 1,800 lithium-ion batteries in the field, is tapping this technology. The feedback is good. "The soldiers are happy because they are getting longer duration times," Ms. Zanardelli said. The batteries are rugged. "They dropped one out of an airplane and it was still intact when it landed," she said.

The DOD also is trying to improve the power and energy density of lithium-ion batteries by investing in alternative chemistries, new materials, and thermal management research. Ms. Zanardelli said the military has been investing in research and development programs over the past six years to improve Silent Watch, Silent Mobility, hybrid-electric boost power, power for soldier communications, and pulse power for armor, among other things. TARDEC is focusing on ultra high-power battery technology using thin electrodes, she said. It also is researching electromagnetic armor and high power density for survivability technologies.

In all, 60 TARDEC energy-storage R&D projects are underway, Ms. Zanardelli said. They are classified into basic research, applications, manufacturing processes, battery management and safety, and alternative systems. There are cooperative research agreements and SBIR programs.

---

[36] The U.S. Army ManTech program funds "development of low-risk, affordable technologies for the military. The goal is to improve production cost through product flow improvements, simplifying product design, and reducing labor hours." Source: http://www.armymantech.com/overview.html

One key basic research project studies the mechanism of "thermal runaway" in valve-regulated lead-acid batteries and is trying to find ways to suppress it. Thermal runaway can cause batteries to overheat and even explode. Currently, such lead-acid batteries are the most widely used in the field, and the Army wants to be able to keep them in vehicles longer, Ms. Zanardelli said.

Among the applied research projects are lithium-ion batteries for Silent Watch and starting, lighting, ignition systems that would replace lead acid batteries. TARDEC is looking at large-format lithium-ion phosphate and nickel cobalt oxide batteries that are lighter, run longer, and offer greater temperature range, she said.

Manufacturing research focuses on trying to improve domestic capability for lithium-ion cells and packs, Ms. Zanardelli said. These efforts concentrate on improving the cells for a 28-volt battery pack and the feasibility of module and pack manufacturing, she explained.

Numerous research projects for battery management and safety are funded in-house and through the SBIR program, she said. They evaluate systems from various manufacturers under cooperative research agreements. A heavy brigade combat team has released battery management specifications for lead-acid batteries in the field.

In terms of alternative systems, TARDEC is conducting research on lithium-titanate hybrid vehicle packs. It also is looking at ultra-capacitors as an option for starting vehicles in conjunction with a battery for Silent Watch, she said.

In summary, Ms. Zannardelli said TARDEC is working with a diversified base of customers and is "actively seeking collaboration" with other government agencies, military contractors, and battery partners.

THE KENTUCKY-ARGONNE NATIONAL BATTERY MANUFACTURING R&D CENTER

Ralph C. Brodd
Kentucky-Argonne National Battery Manufacturing R&D Center

Establishment of the new Kentucky-Argonne Battery Manufacturing Research and Development Center is "a new activity that has been a challenging one and very important for the future of our country," said Dr. Brodd, the center's director and president of the consulting firm Broddarp.

The University of Kentucky has had a research center for around for 30 years that mainly focuses on carbon materials in coal, Dr. Brodd explained. The center already has been working on ultra-capacitors and been very successful at developing new carbon materials, he said. It also has recently been working with new materials such as grass.

The university recently received $10 million for the new center from the National Institute of Standards and Technology under the Recovery Act and $4 million from the Kentucky state government to build a 36,000-square-foot laboratory for advanced batteries. "We will have a reasonable size laboratory and staff," he said. Partners are Argonne National Laboratory, the University of Kentucky, and the University of Louisville. "They have all agreed to work together, which is going to be a very interesting exercise," Dr. Brodd said. "Each has a unique identity. But everyone I have been talking to has been very supportive and has been really interested in seeing the center come to fruition."

The center's initial role was to assist Argonne fabricate state-of-the-art lithium ion cells and to qualify the performance of new cell chemistries, Dr. Brodd explained. But the role broadened. It was determined the industry needed a facility to do more complete development work. "If you have a laboratory coin cell[37] and try to tell somebody that you have a brand-new material that has these wonderful properties, they will say, 'Yeah, sure,'" he said. "They know from experience that a coin cell is not always adequate to define a new material. They really want some real-life cells they can use to quantify their performance and see if these new materials will really measure up."

The center also will develop manufacturing lines. "Up until now, we really have been relying on foreign sources for all of our production equipment," Dr. Brodd said. "There need to be new concepts and new processes to produce batteries more efficiently and at lower cost. The interesting part for me is seeing an engineering center develop around making lithium ion cells, not just the cylindrical cells you have in your laptops today but things that can power vehicles around the country."

Kentucky can be cost-competitive in manufacturing such cells for the world market, Dr. Brodd maintained. "In a sense, Kentucky is going to compete a little bit with Michigan," he said. The space is large enough, however, that "there is no question we can both serve very effectively."

Dr. Brodd said his personal goal is "to re-establish the United States as a world leader, not only in materials and development but in manufacturing technology and capability." He said he has "spent a long time thinking about how to make sure we understand the cost structures and effectively manage the system."

The first goal is to accelerate production of advanced systems and new technologies from national laboratories or universities. "We will be prepared to protect proprietary interests if you have a material you say is unique. We will hold that in strict confidence," Mr. Broad said. "I think that is one of the key issues when you begin to be a service center."

The Kentucky-Argonne center intends to generate new opportunities "not only for Kentucky but also for other local and national entities," Mr. Broad said. It aims to help "close the gap in materials development and cell manufacturing,"

---

[37] Coin cells refer to a small, round battery formats commonly found in consumer electronics.

he said. “The two need to be together in order for any production process to be successful.”

The center wants to facilitate interactions with industry, universities, and national laboratories to “optimize a good supply chain and develop a viable battery manufacturing industry here in the U.S,” Dr. Brodd said. “We are looking for competitiveness worldwide. I am convinced we can do it if we really want to do it. There is no question in my mind.”

The center will design new cell manufacturing concepts in both the cylindrical and prismatic formats, Dr. Brodd said. It also hopes to engage in new cell designs and fabrication processes that give more efficiency and higher performance in terms of speed, density, and cycle life.

Until now, cycle lives of lithium-ion batteries for notebooks have been very short. One thousand cycles has been deemed plenty. “That doesn’t measure up any more,” he said. “All of the auto companies are talking about 100,000 to 500,000 charge and discharge cycles. It is a new game. The designs of cells have to follow the requirements of the industry.”

The center would like to serve as a “focal point” for new battery development in Kentucky and on a national basis, he said. Construction of the building is expected to begin September 2010 and be ready for occupation in 2011.

A roadmap for advanced batteries also is being developed. The project looks at what technologies are available and industry demands for certain vehicles so that cells can be designed properly. “We are looking at the roadmap to be an ongoing situation, something we would renew every two or three years,” Dr. Brodd said. “As you all know, it has been a very dynamic market in these last several years.”

The roadmap also would identify the infrastructure and technology elements “that are required to develop and maintain a leadership position that we feel we absolutely must generate,” Dr. Brodd said. “We really have to re-establish the United States, not just as a technology interest but as a product interest. The country must have that to succeed.”

DISCUSSION

The moderator, Dr. Wessner, asked Mr. Howell to pretend he were in a Congressional hearing. He asked how Mr. Howell would respond if he were asked: “Isn’t this like picking winners and losers? Shouldn’t you let the market do this?” Dr. Wessner also asked what were the comparable international efforts and how “we stack up against our colleagues and competitors around the world.”

Regarding picking winners and losers, Mr. Howell responded that the Vehicle Technologies Program awards R&D grants in a competitive fashion. “It is the typical process,” he said. “Even with the USABC, we put out a request for initial proposals for a certain area, explain what our requirements are for successful proposals, assemble a team to evaluate proposals, and make a selection." Another factor is that the electric-vehicle battery community is fairly small, Mr. Howell said.

The DOE also is "trying open up opportunities for new ideas, even in the research area," Mr. Howell said. The exploratory research program puts out requests for proposals in a new area each year. In 2009, the topic was electrolytes. In 2010, it was new anode materials. In 2011, the topic will be new cathode materials, he said. The program circulates funding announcements through the usual contracting agencies at the DoE, he said.

Mr. Howell said he plans to issue a major Funding Opportunity Announcement in 2010. "This will be a significant opportunity for industry," he said. It will target three areas. One topic for batteries will be next-generation chemistries at the cell level approaching 300 kilowatt hours per kilogram, he said. Another will be for low-cost technologies, such as ultra-capacitors, "anything that can drive the cost down in these batteries," he said. A third topic is design optimization, such as packaging or thermal management technologies that can cut packaging cost.

Over the next few years, Mr. Howell said he is looking to put $45 million of DOE funds into such research. "So there is quite a bit of opportunity for industry to participate in free and open competition."

In terms of international competition, Mr. Howell noted that Japan obviously is putting a lot of R&D money into lithium-ion batteries for vehicles as well as the grid. The Chinese "are coming on strong, not necessarily with a lot of R&D, but with a lot of start-up companies and government support for commercialization of technologies," he said.

The Koreans have "some major players" in the battery industry "and the government certainly puts in money to support that industry," Mr. Howell said. In Europe, "there have been spots of money here and there for research" in electrochemistry, he said. "The Europeans are starting to come on strong in battery development, recognizing how competitive it's going to be over the next decade." The Germans and French in particular are investing aggressively.

Dr. Wessner commented that all of these issues should be explored further.

An audience member asked Mr. Howell if there is a timeline for releasing the $45 million funding announcement.

The Recovery Act "has kind of swamped our funding agents," Mr. Howell replied. "If I don't have Recovery Act on my first slide then they aren't going to talk to me this fiscal year." He said Recovery Act funds must be awarded by September 30, 2010. He said he hoped to publish the funding announcement in late September. "It is important to a lot more than just the battery program," Mr. Howell said. "Other than that, I really can't give you any concrete timeframe or date the announcement will be made. But be prepared. I already told you what is in it, and I'm not planning to change that."

Larry Drzal of Michigan State asked Mr. Howell to explain the $34 million for FY 2011 for a "battery and energy storage hub."

The project is under the DoE's Office of Science, Mr. Howell explained, and is similar in concept to an Energy Frontier Research Center, "but obviously a lot larger than that." He described it as a program "to develop a center or more than one center of R&D in the nation focused on basic electrochemical research."

To get a better idea of the concept, Mr. Howell suggested looking at several hubs already awarded by the Office of Science. "I don't expect the structure to be any different than what they've awarded in the past," he said. The President requested that the center focus on energy storage, but he noted that "it is not obvious that this is going to happen at this point."

Dr. Wessner asked Ms. Zarnadelli of TARDEC whether the Small Business Research Innovation program has been an effective tool for the U.S. Army and whether it provides enough money.

SBIR projects do have enough money, she responded. She said TARDEC has been very successful in getting grants in the battery area, with three topics funded in 2009 and another three approved in 2010. "So we are very successful in SBIR topics and getting Phase I and Phase II funds," Ms. Zarnadelli said.

Dr. Wessner asked whether there was much pressure for the Army to reduce requirements of battery systems.

"No," Ms. Zarndelli responded. Assessments are being done on the power of mission equipment, but "some are just necessary," she said. There also is a new requirement for environmental control, which means air conditioning to keep crews cool. "That and the load needed to cool power electronics are what really are driving these power requirements up," she said.

Dr. Wessner asked Dr. Brodd of the Kentucky-Argonne Battery Center whether there are "substantial industry partners willing to contribute" to the center. He also asked whether Kentucky would be able to maintain funding given the difficult economic climate and prevailing "short-termism" in the country.

The state of Kentucky can sustain the program, Dr. Brodd replied. The state has agreed to fund the center, and the deans of engineering of Kentucky and Louisville have strongly endorsed it and are enthusiastic about it. "But as things get going it will require help with equipment and things of that nature," he said, adding that the center probably will be calling on the generosity of David Howell at the DoE.

Panel VI-B

Federal and Michigan Programs to Support the Battery Industry

Moderator:
Sujai Shivakumar
The National Academies

THE DEPARTMENT OF COMMERCE AND THE ROLE OF THE MANUFACTURING EXTENSION PARTNERSHIP

David C. Stieren
Manufacturing Extension Partnership

The Manufacturing Extension Partnership "has a programmatic focus and a base of resources that I think is very relevant to addressing challenges as the battery industry for vehicles develops in this country," said Mr. Stieren, who oversees technology deployment through the program's nationwide network of 66 centers and 460 service centers.

The MEP is a "federal-state-private partnership" managed at the National Institute of Standards and Technology in the Department of Commerce, Mr. Stieren explained. NIST has programs that work on research, performance characterization, and measurement methods for battery technologies. NIST also administers the Technology Innovation Program, which awards grants in the battery sector, he noted.

The MEP "basically exists to provide assistance to the nation's manufacturers," Mr. Stieren said. "We provide services and access to public and private resources targeting enhanced growth, improvement in productivity, and enhanced capacity."

The program was created in the late 1980s, when NIST changed its name from the National Bureau of Standards. "We work with companies that want to be proactive, want to expand, and want to establish their niche in the marketplace," he said. "We also are a program that emphasizes the services we provide to manufacturers." The programs are measured in terms of the economic impact to clients the MEP serves, he said.

Services are delivered through MEP's network of 60 centers, which are found in each state and have 1,600 staff. The MEP also operates 440 field applications, Mr. Stieren noted. The staff are "boots on the street interacting with manufacturers on a daily basis to provide business and technical assistance," he said. Because client bases tend to differ, not all centers offer the same services, he explained. The MEP also contracts with 2,300 service providers. "If a center doesn't have the capability for a specific need, they can go into our network to look for the capability or contract with third-party service providers," he said.

Most companies that manufacture products in the U.S. are within a two-hour drive of either an MEP center or assistance service location, Mr. Stieren said. "We really have a fantastic reach to the nation's manufacturing base," he said. Each year, the MEP interacts with some 31,000 companies. "That is a high number and a good number, but there are plenty of opportunities for us," he said. Currently, the program works with less than 10 percent of America's 340,000 manufacturers. "We're always looking to increase those numbers," he said.

Partnerships are the real strength of MEP programs, Mr. Stieren emphasized. "One of the things our network is really good at is helping companies get access to resources and capabilities they may not know about or have the wherewithal to gain on their own," he said. MEP advisors work to connect companies to state and federal resources.

MEP advisors also help companies gain access to needed technologies, Mr. Stieren said. The MEP can connect them to technology at national, universities, and even private labs, he said. It also works with the Labor Department and other partners on workforce-training issues.

The MEP has many kinds of partnerships. At the federal level, the MEP works with agencies such as the Energy, Defense, and Labor departments, the Small Business Administration, and the Environmental Protection Agency, Mr. Stieren explained.

Each MEP center also partners with state agencies. "Our centers operate as not-for-profits and in some cases as components of state agencies or university affiliates," he explained. The MEP's approach to partnership "depends on the specific state and the specific need we are trying to address," he said. Each of the MEP's 60 centers have cooperative agreements, with states providing matching funds. The MEP also works with many state science and technology institutes, Mr. Stieren added.

The program looks across a state to understand all of the technology-based economic activity. It also frequently partners with state economic development organizations "so that we can better align what our centers are doing and the services we are offering across our system on a national basis with the needs and interests of specific states," he said.

There also are partnerships with trade associations. "This is a good way for us to get a feel on a broad basis for where industry sectors are going so that we can direct the focus of our centers and services," Mr. Stieren said.

The MEP's economic impact has been impressive, Mr. Stieren said. In 2009, its programs helped generate $9.1 billion in new and retained sales and $1.7 billion in new investment by clients. They also led to $1.4 billion in cost savings and created or helped retain nearly 53,000 jobs. He said these data were obtained through an independent, third-party survey of more than 7,000 companies that completed projects in 2008. "This is just a snapshot of the types of impacts we have across this program," he said. These results were achieved with $330 million in state and federal funding and fees charged to clients in industry. "These are pretty impressive impact numbers for that kind of investment," he said.

For the U.S. battery industry, the MEP conducted around 120 projects with companies around the country between 2005 and 2009, Mr. Stieren said. The projects involved 47 different companies in 26 states. Roughly one-third of those companies had 50 employees or fewer. About half had more than 100 employees, meaning they are "pretty established in manufacturing," he said. "This is pretty indicative of the type of clients were serve in the MEP," he said.

The challenges addressed by MEP programs "were across the board," he said. They include Six Sigma quality, marketing, road-mapping, lean manufacturing, energy efficiency, export market access, supply-chain management, and product development, he said. These battery projects are credited with helping generate $69 million in sales, $35 million in cost savings, $32 million in investment, and 1,041 new or retained jobs. In 2009 alone, the MEP worked with 12 different battery projects credited with generating $8.6 million in sales.

A new MEP strategy for aiding manufacturers has implications for the battery industry. In its first 20 years or so, "MEP was very focused on product efficiencies and continuous improvement of helping U.S. manufacturing companies compete," he said. "MEP now also is very focused on growth."

The MEP commissioned a survey asking client companies what they saw as their three most strategic challenges over the coming three years. The survey found that "continuous improvement" was seen as manufacturers' biggest challenge, cited by more than 70 percent of respondents. Finding growth opportunities and product innovation also were cited heavily. The MEP tells companies "that if they interact with us they can continue to improve their bottom line and reach their goal of cutting costs by 20 percent," Mr. Stieren said. "They also can look for opportunities to grow. It is not something set in stone. It is a target."

The program's five key focus areas still include continuous improvement, which Mr. Stieren described as "a given" because "you have to be lean, have quality in place, and address that on a continuous basis." Other top priorities are technology acceleration, supplier development, sustainability, and workforce issues.

The MEP advises companies on how to turn their technologies into processes, products, and services needed to bridge the Valley of Death, Mr. Stieren explained. "Our approach is to make the connection between companies and their needs to the sources of technology that are out there," he said. "That can mean federal, university, and private laboratories. Very frequently, small technology-based start-ups just don't have the access and wherewithal to continue to evolve their technology along the maturation path and get access to the technologies they need to scale up."

The MEP's supply-chain work also is becoming more important. Around 68 percent of U.S. manufacturers operate as part of a supply chain, Mr. Stieren noted. "This is interesting, because it means that most of our manufacturers' products go to other manufacturers, not necessarily to an end consumer," he said. "So for a national program targeting assistance for manufacturers, it is important that we understand what operation within a supply chain means."

The MEP strategy works with suppliers to help them understand their niche in the supply chain, he said. The MEP also works with OEMs and federal agencies "so that there is a top-down and bottom-up approach to help the overall manufacturing supply chain, not just a company within it," Mr. Stieren said.

One way Mr. Stieren said the MEP may provide value to the battery industry is by linking it up to good companies that may be in other manufacturing sectors but that are now looking for business. "We have seen this in our partnerships with a lot of other industries," he said. In Michigan, the MEP is helping companies assess their capabilities and diversify their market. "We have access to a lot of companies that may be relevant to the supply base for the battery and electric-vehicle industries as they develop," Mr. Stieren said.

Supplier-scouting is another MEP activity that may help the battery industry. The MEP works with OEMs and federal agencies with very specific needs to find companies with the right technical capability or production capacity to supply them, Mr. Stieren explained. "Through our national network, we basically go out and find the companies that meet their needs," he said.

The MEP is working with federal agencies and manufacturers of wind-power generation equipment, nuclear equipment, and military products, for example. The MEP is "doing some pretty interesting things" with the DOE connected to Buy America goals of projects funded by the Recovery Act, he said. "Really what this is all about is leveraging the knowledge that resides across our MEP system to pre-qualify manufacturers and deliver capable suppliers to those OEMs."

The MEP is a proven manufacturing-assistance program, "and we have a great reach and direct connection to U.S. manufacturers," Mr. Stieren said. "We think there is really significant potential to leverage our national network and our existing partnerships, including those at the state level, to further develop the U.S. battery industry for electrified vehicles."

MICHIGAN INVESTMENTS IN BATTERIES AND ELECTRIC VEHICLES

Eric Shreffler
Michigan Economic Development Corporation

Michigan's aggressive push into batteries began with a thorough study of its economy launched three years ago, explained Eric Shreffler, who leads the MEDC's advanced energy storage program. "As many of you know, Michigan has been experiencing very difficult times," he said. It realized that it must diversify its economy beyond auto manufacturing.

State economic development officials "looked at growing industries, industries that needed a little boost to get over the hump, and best practices around the globe," Mr. Shreffler explained. In particular, they were intrigued by the Swedish "triple helix" strategy for developing industrial clusters with public-private partnerships that connect government funding, investment from private industry, and research programs at universities and national laboratories,[38] he said. It also looked at nurturing clusters with key anchor companies in an industry "that can set the stage for growth of the supply chain and customer base."

The MEDC formed teams for specific clusters related to renewable energy. Mr. Shreffler headed the advanced energy storage team. Other teams focus on materials, bio-energy, solar photo-voltaic cells and panels, water technology, and wind. "All of us are looking at crafting strategies to develop those sectors," he said. In the battery sector, for example, Mr. Shreffler's team saw a need for new investment incentives and legislation.

Advanced batteries seemed logical because the market is projected to reach $20 billion in annual revenue in 2020. The industry also fit Michigan's strength in auto manufacturing. Although the state is more dependent on that sector than any other, Mr. Shreffler said "that doesn't necessarily mean we should diversify completely away from that. There is an opportunity to diversify within the automotive industry. The move for vehicle electrification has basically crossed the tipping point. This is a real thing. It will become part of the portfolio of the OEMs."

Batteries are strategically important because they will be "the new power train" of future automobiles, Mr. Shreffler said. "Michigan did not want to stand by and cede leadership in power-train development to other states and countries."

---

[38] Triple Helix in the study of knowledge-based innovation systems refers to interaction among universities, industry, and government. The Triple Helix concept has been championed by Henry Etzowitz. See *Triple Helix: A New Model of Innovation*, Stockholm: SNS Press, 2005 (in Swedish) *The Triple Helix: University-Industry-Government Innovation in Action*, New York: Routledge, 2008

Part of Michigan's strategy was "to seed the marketplace and send a signal we are serious about developing this ecosystem here in this state," Mr. Shreffler explained. "We wanted to be the first in North America to really push that message through." As a first step, the state offered strong incentives to lure anchor companies in the sector.

Michigan was fortunate to establish its incentive program before passage of the Recovery Act made billions of federal dollars available for the advanced battery industry. "We did not know at the time we were doing this that there was going to be a Recovery Act," Mr. Shreffler said. "But we felt pretty confident there were going to be significant investments at the federal level in advanced battery technology. So we wanted to make sure that as we were designing incentives that companies could leverage our incentives as cost-sharing for those to-be-determined federal opportunities." Some federal tax credits were available that could dovetail with Michigan's, such as those for consumers purchasing vehicles.

Once the battery companies came, the state worked with them to diversify beyond transportation and into other complementary markets that could use their battery products, Mr. Shreffler said. The state also focused on the entire supply chain for batteries.

The MEDC began by targeting "the heart of the value chain" for batteries—the cell and battery-pack factories and vehicle electrification programs of major auto makers. "We wanted to solidify and cement as much of that here in Michigan as possible," Mr. Shreffler said. The MEDC saw a need for "very aggressive incentives."

Centers of Energy Excellence[39] are a key part of this strategy. This program "was developed to allow the MEDC for the first time to provide direct grant money to for-profit companies," he said. These companies would be leveraging outside federal financing and partnering with universities or federal labs to commercialize technology. The first six grants worth $43 million were made in 2008, with $13 million going to Sakti3 to develop solid-state batteries and to A123 Systems to establish a pilot assembly facility. The other $30 million went to bio-energy projects.

The other major action was the Michigan Advanced Battery Tax Credits (MABC) program,[40] "the first such program in the nation and a very aggressive approach," Mr. Shreffler said. The initial legislation provided for $335 million. It was boosted to $1.02 billion. "What happened is that after we made the announcement, the quality of opportunity and interest from significant, global,

---

[39] Michigan's Centers of Energy Excellence Program was established under Senate Bill 1380, Public Act 175. In the program's first phase, the Michigan Strategic Fund Board awarded $43 million in grants in 2008. For-profit companies receiving grants must secure matching federal funds and financial backing. Public Act 144 of 2009 allowed a second phase of the COEE program.

[40] Michigan's Advanced Battery Tax Credits initiative was created through an amendment to the Michigan Business Tax Act, Public Act 36 of 2007, to allow the Michigan Economic Development Authority to tax credits for battery pack engineering and assembly, vehicle engineering, advanced battery technology development, and battery cell manufacturing.

leading battery companies was such that we were able to go back to the legislature and show them this was not a build-it-and-hope-they-will-come" gamble, he said. The MEDC showed "if the state offered these incentives, these companies will come and establish a presence in Michigan," Mr. Shreffler said. "I think I can count on one hand the number of 'no' votes across both chambers every time we went back and discussed the opportunity with the legislature."

The credits cover four main areas. Of the total, $255 million went to battery pack manufacturers, who receive tax credits for every pack they assemble in Michigan. "The larger the battery, they more the credit, and the more you do the more you get," Mr. Shreffler explained. Another $135 million went to vehicle engineering and $30 million to advanced battery technology projects. Most of these credits went to auto manufacturers. All recipients "are global companies that could utilize R&D and engineering resources around the globe," he explained. "We wanted them to really establish their engineering and development work for advanced vehicle batteries here in Michigan." The tax credits helped offset some of the costs.

The rest of the credits, worth $600 million, went to battery cell manufacturers. Michigan refunds half of the capital investment up to $100 million for a fully-integrated battery cell manufacturing facility. That means everything from the coating processes to assembly, "raw materials in one end, cells out the other," Mr. Shreffler said.

The fact that the credits are "performance-based" helps assure legislators, he said. The credits are spread out over four years, and manufacturers can start receiving the credits in 2012 after they have built a facility and hired people. "If you don't do that, you don't get the credit," Mr. Shreffler explained. "That return on investment by deferring the payout is what ultimately made the legislature feel comfortable that this was the thing to do, combined with the fact that we had considerable opportunities at the federal level as well as with the Recovery Act."

Credits of $100 million each went to six cell-manufacturing plants, those of Johnson Controls-Saft, LG Chem/Compact Power, A123, Dow-Kokam, Fortu Powercell, and Xtreme Power. Credits for pack manufacturing went to GM, Ford, Johnson Controls-Saft, and Dow-Kokam. Battery and vehicle engineering credits went to GM, Ford, and Chrysler. The $1.3 billion in grants through the Recovery Act mainly went to these same companies.

The credits and grants were so popular that from Aug. 5, 2008, "our attraction efforts were just basically answering our telephone," Mr. Shreffler recalled. "So many companies now saw that the center of gravity in North America for this development and manufacturing was going to be taking place in Michigan. They really wanted to come and be part of that."

Since November 2008, investments worth $5.7 billion have been announced in Michigan. There are 16 different projects, "and our pipeline continues to be very full," Mr. Shreffler said. "I suspect in the fall we will be rolling out additional announcements that will be supporting this industry." Michigan battery supply-chain investments now stretch from the Detroit area to Lake

FIGURE 16 Michigan's energy storage industry: supply chain investments.
SOURCE: Eric Shreffler, Presentation at July 26-27, 2010 National Academies Symposium on "Building the U.S. Battery Industry for Electric Drive Vehicles: Progress, Challenges, and Opportunities."

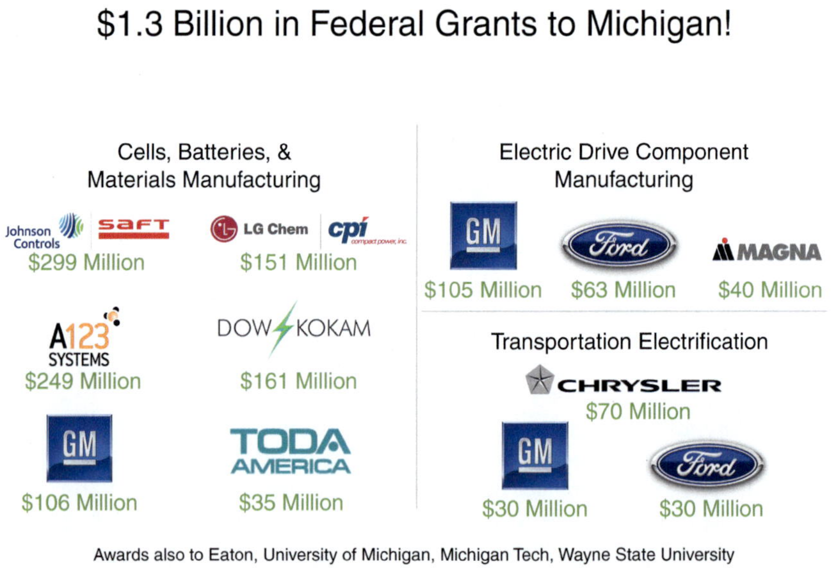

FIGURE 17 Michigan's energy storage industry: federal grants.
SOURCE: Eric Shreffler, Presentation at July 26-27, 2010 National Academies Symposium on "Building the U.S. Battery Industry for Electric Drive Vehicles: Progress, Challenges, and Opportunities."

Michigan. They includes a cathode materials plant by Toda America, battery testing facilities by AVL and A&D Technology, electric motor components by Magna, energy-storage solutions by Xtreme Power, and electric drive-train testing by Eaton.

Many small Michigan companies now also belong to this supply chain. "There are a number of suppliers in this state actively getting contracts and engaged in this industry that we're not even aware of," Mr. Shreffler said. "We have to go to conferences in Florida and California to meet with them and find out they are doing this." Going forward, the MEDC team will start focusing on "understanding the full depth and breadth of what is happening with all of the companies here," he said. It also will try to "build out this eco-system with Michigan-based companies."

The energy storage drive is important for a number of reasons, Mr. Shreffler said. It is establishing a domestic source for high-energy, high-power batteries needed by the U.S. transportation industry and Department of Defense. Having the center of gravity in the U.S. also spurs new economic expansion and development opportunities from transportation-focused companies and "creates a foundation for building out a sustainable ecosystem over time," he said. It also provides further opportunities for collaboration between private industry and Michigan universities.

Now the MEDC is developing the next steps to advance the battery industry. Mr. Shreffler recalled attending a board meeting where incentives were being discussed. An executive asked: "Now that you have taken care of batteries, what are you going to do next?" In reality, "we are really viewing this is as just the beginning," Mr. Shreffler said. "The hard work is yet to come. The easy part was working with great companies to establish opportunities here."

One focus is to look at "dual-use" opportunities for companies in the battery industry. The military is a major potential market for Michigan manufacturers. Energy storage for the grid is another major focus. "All of this is to the end of trying to help our cell manufacturers have long-term sustainability and begin to rapidly drive down the cost curve so that the battery technology is accessible to where it is needed," Mr. Shreffler said.

The MEDC will continue to reassess its "economic tool kit" because the industry is in a new phase, he said. What's more, two-thirds of Michigan's legislature is set to turn over and a new governor will be elected. The MEDC will work to educate new officials on what has been done, what needs to be done, and what further tools are needed, he said. It also will continue to advocate for strong policy at the federal level to make sure the industry keeps moving forward. "It all is about execution," Mr. Shreffler said. "We have to execute as an economic development agency. Our cell manufacturers and suppliers must execute to build out their capacity. The federal government has to execute on not abandoning the path that we've gone down."

One next step is to strengthen what the MEDC calls "the alliance." The goal is to align state initiatives, especially in batteries and advanced materials, with the priorities of federal agencies such as the Energy and Defense departments and the national laboratories. "We look at this as an opportunity for Michigan companies to plug into federal opportunities they may have not had before," Mr. Shreffler said. That could retain jobs in the state and advance common research and commercialization to "help drive this dual-use holy grail we have been talking about," he said.

The state already has entered into a number of cooperative agreements. Michigan, Oak Ridge National Laboratories, and TARDEC have pooled resources for a $27 million development project over three years to commercialize advanced-storage and light-weight material research in DOE labs and adapt it for military use, for example. By demonstrating the approach is successful, "we hope this could be an opportunity for more solid and expanded funding going forward," he said. "It is our charge to really make this a relevant opportunity for all stakeholders involved at the state and federal level to develop dual-use technologies."

While "you will not find our team posing in front of a banner saying 'Mission Accomplished,'" the MEDC has attained some of its early goals, Mr. Shreffler said. It has indentified opportunities, developed a strategy and "unique attraction tools," successfully competed for Recovery Act funds, and begun attracting the advanced battery supply chain, he noted. Now the MEDC is focusing on building out the supply chain, assisting on regulatory and policy issues, and using the alliance to build collaboration among government, universities, and industry, he said. The MEDC also aims to keep exposing companies to complementary markets and pursuing new federal funding opportunities.

Roundtable:

What Have We Learned and Next Steps

Moderator:
Mary Good
University of Arkansas at Little Rock

Moderator Mary Good asked participants to offer thoughts on "where we are and what are the pieces we should pick up with the next meeting that will be devoted to this area." Bill Harris of Science Foundation Arizona, Les Alexander of A123, and Gary Krause of the Michigan Economic Development Corporation were asked to open the discussion.

As "a citizen concerned with the competitiveness of the country and some of the opportunities that clearly exist," Mr. Harris said he was "extraordinarily impressed with what is going on in Michigan relative to the impression that newspapers give of the state of dire despair." He recalled an old GM advertising line, "It's Not Your Father's Oldsmobile," in the 1980s. "I think this is not the Michigan that your father knew as well," he said.

Michigan's outstanding universities and talent pool have long been assets, Mr. Harris noted. "But the tragedy was the narrowness of the economic base for so long around one industry that may not have modernized the way it should have." He said he also was impressed at how well Michigan's government functions. "When I hear there was a vote for $1 billion in incentives and there were a handful of negative votes, I would like to exchange legislatures for a few days" with that of Arizona, he said. "I can get you a whole lot of 'no's' that you seem to be missing." In the upcoming Washington meeting on batteries, "I think it is important to convey the sense of momentum, determination, and I would say competence that is here," Mr. Harris said.

Another interesting message from the conference is the possibility of interactions among states or a big region, such as between Kentucky and Michigan, Mr. Harris said. "I think Kentucky's goals and ambitions with Argonne match nicely with what is going on in Michigan, and there could be reasons to look at doing things together."

It also is important for the U.S. to find a way to do large-scale manufacturing, Mr. Harris said. "I think if we stay with small businesses and are not able to compete with the big boys in Asia, we are making a serious mistake," he said. NIST's activity in manufacturing is important, but after 30 years it may be time to fine-tune that program, he added. "That was an important activity when it started," Mr. Harris said. "It may need to take a look at where it is today and where the world is today and be bold about what it can do for the future. I say that not as a criticism, but as a challenge, opportunity, and complement to what it is needed."

Finding better ways for federal agencies to partner with states to add real value to the economy also is important, Mr. Harris said. "Too often, it seems the feds come in with a good idea," he said. "But they are not really connected to the state apparatus and state entities. I also do not think we get the return on investment that we need" he said.

Mr. Harris suggested that the National Academy of Sciences invite speakers of state houses of representatives to these forums. "I think you need some legislators to understand what other states are doing," Mr. Harris said. "The absence of informed representatives hurts the dialogue. Too often we forget about including state legislators in meetings like this, and that hurts their knowledge and information."

Dr. Good said she liked the idea of inviting state representatives. "It is true that those are the fellows who have to vote 'yes' on some of these initiatives, and we need to get them on our side," she said. The real problem, she acknowledged, is getting them to come. But if they are invited to Washington for such a meeting, the state will pay for it and some might see value in such a trip, she suggested.

Dr. Good asked Les Alexander, general manager for government solutions at A123, for his view from an industry perspective.

The coordination of state, federal, and military efforts has been important to driving development of the advanced battery industry, Mr. Alexander said. "Where I hope we continue to go is to look at demand-driven stimulation rather than stimulating manufacturing and research," he said. "That is not to say manufacturing and research aren't important. But we need to continue to pull through the technology into demand and applications to get the vehicles out on the road."

Measures such as federal fleet purchases, incentives to encourage cities to adopt alternative energy, and grid applications "all are important to continue to strengthen the foundation we have built here," Mr. Alexander said. "It is important for legislators and researchers to continue to look at the end game, which is to change our economic environment and fossil-fuel use." While battery companies can keep conducting research in these areas, "it is important to get applications out on the street," he said.

Dr. Wessner asked Mr. Alexander to cite the main risks facing A123.

One of the biggest is that A123 will not be able to operate its new U.S. plants at full capacity, Mr. Alexander said. Although A123 currently projects that its capacity will be fully utilized, if electric vehicles are not built or purchased "there is a risk this industry will go away," he said. Successful launches of the Volt and Leaf are important.

Mr. Alexander noted that other nations that do not have the immense infrastructure of the U.S. for motor vehicles are looking at hybrids and electrics as their main source of transportation—just as cell phones are more important than land-line telephones in certain emerging markets. "If we don't embrace this technology, we could lose it," he said.

Dr. Good agreed. "That means the issue of government procurement becomes very big." She asked Gary Krause of the Michigan Economic Development Corp. for his opinion.

Mr. Krause first commented on the presentation by Eric Shreffler of the MEDC, which made it seem that the whole process was "so orderly and methodical." From the inside, he said, "the exercise was a lot more exciting and a lot more breathtaking at times, with outcomes not necessarily predetermined."

He said he was pleased to hear there was general agreement that the task of policy is not finished. "The easy part is incentivizing companies to get involved in the industry," he said. "The more difficult aspect of this is that, now that the dog has chased and nearly caught the car, what do we do next? That is something that we as a state are really very concerned about." Michigan has "literally bet the farm on this particular aspect of a broader policy for industrial diversification," Mr. Krause said. There is $6 billion in state, federal, and private investment on the table, Mr. Krause pointed out. "That is a lot. So the issue of completing this task from a policy standpoint really is key." The MEDC "will be doing everything in our power to protect, enhance, and leverage those considerable investments that have already been made," he said.

In terms of market drivers, Mr. Krause said he "was a bit disappointed." The National Academies should address "why it is so difficult for federal agencies, which have their hands on certain levers, to incent this industry," he said. "Why is it so difficult to get better cooperation in terms of military applications, the postal service, and even general governmental use of vehicles?" He said response has been "tepid, and we really need to concentrate on that a little more."

Another major issue is acceptability of electric vehicles, Mr. Krause said. "I was encouraged that the discussion got beyond passenger cars," he said. "The real payoff comes when one gets into those larger trucks, the off-road vehicles, and the construction and agriculture applications.

Mr. Krause said Mr. Sperling's presentation on the cultural shift in attitudes toward electric vehicles was interesting. "What is needed here is a very educational process," he said. "This thing is seemingly being pushed from a government and industry standpoint. And all of the discussions seem to go around why we can't do this rather than why we can do this." There are "real advantages" to electrifying transportation, he noted. "That excitement just doesn't seem to be coming through. I would suggest that the educational effort be very different and not have a heavy fingerprint of government on it. It really needs to be about why there is an advantage to electrifying vehicles." Mr. Krause said if the push is from the federal government perspective, "the very shrillness of the atmosphere those kinds of discussions take place in will discredit the process."

Dr. Sastry of the University of Michigan said she "couldn't agree more" with Mr. Krause's point that "this cannot be a top-down, force-fed kind of thing." Instead, "we really need to engage the next generation of companies and people," she said. Dr. Sastry said she also agreed with other speakers that the U.S. still needs to keep stressing innovation and scaling up, and that success stories should be highlighted. She suggested engaging student teams, education programs, and programs like the X Prize.[41] She noted that the University of Michigan's Energy System Engineering program proved successful and that others will follow. "What you are seeing now with Michigan Tech and Wayne State are follow-on programs that are excellent and marks of success," she said. "What Andy Levin talked about was a mark of success." Dr. Sastry said it would be great if a forum were created for innovators in the industry and academics.

Regarding Mr. Krause's concern that too much of the vehicle electrification effort is driven by Washington, Dr. Wessner said that the goal "is to keep the DOE and Congress focused and keep at least some of the funding coming."

Dr. Wessner also noted that Detroit has been able to sell cars of "great to medium quality through advertising for years." He said he loves the Jeep ad campaign "What America Makes Matters" and that it is a great message. "I am quite confident they can sell these cars," he said. He said he understands the Tesla electric car "smokes." Dr. Wessner also recalled that when Detroit pushed sport-utility vehicles, a substantial tax credit was successful. Perhaps the $7,500 tax credit for electric vehicles should be doubled, he said. Dr. Wessner said he also wonders if Michigan can impose a tax on gas and use the proceeds for the advanced battery initiative.

As in many states, gas taxes are "a very hot button right now," Mr. Krause responded. "We haven't figured out what happens as vehicles become more efficient," he said. "As we burn less gas, it decreases the amount of dollars available for infrastructure and road improvements. That is something that has been plaguing many states." Michigan almost left more than $400 million in its share of federal gas revenues coming back to the state because it couldn't match the percentage to access those funds, he explained. "So the issue of how we tax transportation is going to be a critical debate, and Michigan is just one of many states with a similar problem," he said. "So the ability to divert some of those dollars to even higher objectives, I suspect, isn't a priority yet."

Dr. Good said she agrees the "whole question of how to maintain the infrastructure presently paid by gasoline taxes will be interesting as we get greater efficiency. We still have to pay for the infrastructure."

Louis Infante of Ricardo Engineering asked whether there are similar forums to discuss the broader changes required in transportation infrastructure of the future that are needed to support the battery industry. He asked if the National Academies sponsors a forum on the transportation system "that would allow all of us to see how the good work done here and supported by the state of

---

[41] X Prizes of $10 million and up are awarded by the X Prize Foundation, a non-profit institute, for breakthrough accomplishments such as development of the first private vehicle for space and ultra fuel-efficient vehicles.

Michigan integrates into things like the grid." This would reduce business risk and give investors and a better idea how to run their businesses, he said.

Dr. Good responded that the National Academies does not do enough of that. "We have a tendency to look at a piece of the system, but not understand the impact a piece has on the system as a whole. Many times, the piece then fails because we did not properly understand how it would fit into the system," she said. "We just aren't very good in this country at doing systems analysis." The National Academies has done some work on the topic, "but it is fairly academic," she said. "As far as I remember, most have been anchors for doorways."

Dr. Wessner agreed that "this country is not very systemically inclined, even in matters of great import, such as defense." While more should be done, the problem is always making such studies relevant to lawmakers busy with constituents and to Congressional committees, he added. The Transportation Research Board of the National Academies has a "vast range of ongoing work in this area," Dr. Wessner said.

What distinguishes the STEP board "is that we focus on how to actually deploy and what intermediating institutions are needed to bring public-private partnerships together to facilitate the advance of technology," he said. "What you heard from MEP and MEDC is the sweet spot of our analysis—how do you push technologies up the learning curve and down the cost curve." The board also tries to come up with "actionable items" for legislators. "They can't execute on broad analysis," he said. "They need very particular things to do."

Dr. Wessner also explained that this conference is part of a wider study of state and regional policy. The STEP board is looking at best practices in bringing technologies forward. "Needless to say, this is a critical area," he said, "and one that has national implications."

To encourage more focus on macro system issues such as the grid and the transportation system, the debate should be framed in terms of national and economic security, Mr. Harris suggested. "President Eisenhower was able to get the national highway system done that way, and we have not evolved a lot since then," he said. "On national security issues, we are good. And transportation and the grid are clearly national security issues. The longer we kick this ball down the field, the more difficult it is going to be to really change the country and free ourselves of the Middle East."

Legislation that increases demand will help create the infrastructure, Mr. Alexander said. "As a battery company, I cannot go to the party without the vehicles. We can create the best battery in the world, but without vehicles to put them in what will happen is that this industry will go back overseas and we will have stimulated another country's industries."

The government needs to say it will convert half of its fleet and spur buying programs by municipalities, Mr. Alexander said. "We need a national buying program so we can move this technology." In terms of passenger cars, consumers must see electric vehicles on the road to become interested, he said. "Your neighbor has to have one," he said. "That is the best source of advertising.

Then it will take hold. Right now, it's not going to take hold by having five or six cars here and there, or a military base that has a golf cart running around. We appreciate these little pockets of the DOE doing something over here and the DOD doing something over there. But we need a massive, consolidated effort."

Dr. Good said the upcoming Washington meeting on batteries should include people who can speak to "pull-through" and whether or not there is an appetite for getting people to talk about procuring such vehicles. "In many ways, that seems to be the secret of success," she said.

Nothing advances a technology faster than on-the-road experience, Mr. Krause said. "That is why it is so critical to get deployment," he said. Deployment also will help development of manufacturing and the supply chain, and provide knowledge useful to advance everything from basic materials research to grid technology. "But let's stop talking about this stuff and let's get those vehicles, knowing that the first generation isn't going to be without some difficulty," he said.

A member of the audience asked about the influence of fuel prices. "In the industry, we were all amazed to see the kinds of shifts in behaviors that took place just a couple of years ago when fuel prices went high," he said. "We saw people who were absolutely diehard SUV drivers suddenly downsize and were happy about it." Even though fuel prices are not the complete story, "that lever, that ability to regulate what is happening with fuel prices is a really critical part of this puzzle," he said. "Without that you aren't going to get the demand and we don't have a battery industry. I think it deserves a lot of focus."

Dr. Wessner concluded the conference by thanking representatives from the state of Michigan and leading universities for taking part. "You have provided us with a remarkable range of information and insights, which will be extremely helpful," he said. An important role of the National Academies is to convey such information to Congress and the DoE. "When people say U.S. government, states, and the university community cannot cooperate, they clearly have not been in this room. I congratulate you for that."

Dr. Wessner thanked the MEDC and Gary Krause in particular for their help in organizing the conference and lining up speakers.

III

APPENDIXES

Appendix A

Agenda

Building the U.S. Battery Industry for Electric Drive Vehicles: Progress, Challenges, and Opportunities

A Symposium Organized by
The U.S. National Academy of Sciences
in cooperation with
The Michigan Economic Development Corporation

Agenda

26-27 July 2010

Schoolcraft College VisTaTech Center
18600 Haggerty Road
Livonia, Michigan

Day 1: Monday, 26 July 2010

9:00AM **Welcome**
Greg Main, Michigan Economic Development Corporation

Opening Remarks
The Honorable Carl Levin, U.S. Senate (D-Michigan)
Sridhar Kota, White House Office of Science and Technology Policy
The Honorable Jennifer Granholm, State of Michigan (via video)

9:40AM **Overview of NAS Study: Building the Battery Industry for Electric Vehicles**
Mary Good, University of Arkansas at Little Rock

9:50AM **Panel I: The Federal Outlook for the U.S. Battery Industry**
Moderator: Charles Wessner, The National Academies

The Department of Energy Perspective
Patrick Davis, U.S. Department of Energy

The Army Perspectives
John Pellegrino, U.S. Army Research Laboratory
Grace Bochenek, U.S. Army Tank and Automotive Research, Development, and Engineering Center

10:45AM **Coffee Break**

11:00AM **Panel II: The State of Battery R&D and Manufacturing in the U.S.**
Moderator: Ralph Brodd, Kentucky-Argonne National Battery Manufacturing R&D Center

The Battery Industry Perspective
Jason Forcier, A123
Mohamed Alamgir, Compact Power

The Automotive Industry Perspective
Nancy Gioia, Ford

The University Perspective
Ann Marie Sastry, University of Michigan and Sakti3

12:30PM **Luncheon Address**
The Honorable Debbie Stabenow, U.S. Senate (D-Michigan)

1:30PM **Panel III: Strengthening the Supply Chain**
Moderator: Jim Greenberger, National Alliance for Advanced Technology Batteries

Defining the Supply Chain: Gaps and Opportunities
Michael Reed, Magna

Battery Manufacturer Perspective
Tom Watson, Johnson Controls

Battery Materials Availability and Recycling
Linda Gaines, Argonne National Lab

2:30PM **Coffee Break**

2:45PM **Panel IV: Market Drivers: Creating Demand for Electric Vehicles**
Moderator: Robert Kruse, EV Consulting

Incentives for the Electric Vehicle Market
Daniel Sperling, University of California, Davis

The Industry Perspective: Transforming the Automotive Industry
Gary Smyth, General Motors

Early Adoption of Hybrid Vehicles
Bill Van Amburg, CALSTART

4:00PM **Panel V: Building the Battery Workforce**
Moderator: Bill Harris, Science Foundation Arizona

Workforce Needs and Opportunities
Robert Kamischke, EnerDel

Technical Training and Workforce Development
Simon Ng, Wayne State University

Day 2 – Tuesday, 27 July 2010

9:00AM **Welcome and Introduction**
Andy Levin, Michigan Department of Energy, Labor, and Economic Growth
Mary Good, University of Arkansas at Little Rock

9:15AM **Panel VI, A: Federal and State Programs to Support the Battery Industry**
Moderator: Charles Wessner, The National Academies

The Department of Energy Battery R&D Program and Goals
David Howell, U.S. Department of Energy

Department of Defense Battery R&D Programs and Goals
Sonya Zanardelli, U.S. Army Tank and Automotive Research, Development, and Engineering Center

The Kentucky-Argonne National Battery Manufacturing R&D Center
Ralph Brodd, Kentucky-Argonne National Battery Manufacturing R&D Center

10:30AM **Coffee Break**

10:45AM **Panel VI, B: Federal and State Programs to Support the Battery Industry**
Moderator: Sujai Shivakumar, The National Academies

The Department of Commerce and the Role of the Manufacturing Extension Partnership
David Stieren, Manufacturing Extension Partnership

Michigan Investments in Batteries and Electric Vehicles
Eric Shreffler, MEDC

11:30AM **Roundtable: What Have We Learned and Next Steps**
Moderator: Mary Good, University of Arkansas at Little Rock

Bill Harris, Science Foundation Arizona
Les Alexander, A123
Gary Krause, Michigan Economic Development Corporation

Appendix B

Biographies of Speakers[*]

GRACE BOCHENEK

Dr. Grace M. Bochenek was appointed Director of the U.S. Army's Research, Development and Engineering Command's (RDECOM's) Tank Automotive Research, Development and Engineering Center (TARDEC) in August 2006. TARDEC is located Warren, MI, and is recognized as the ground vehicle center of excellence and the premier laboratory for advanced military automotive technology for ground vehicle systems and logistics support equipment. A recipient of the Meritorious Executive Presidential Rank Award in 2008, Dr. Bochenek brings over 23 years of scientific, technical and managerial experience to this preeminent Army institution. She creates and leads all research, development and engineering strategies for the Department of Defense's Ground Vehicle Manned and Unmanned systems with military impact worldwide. In this role, she oversees an annual budget of over $500 million in research and development funds as part of the annual $24 billion Department of Army Ground Vehicle and Logistical System investment plan. She manages a workforce of over 1,200 government civilian, military and contractor employees and sets strategic direction for a full range of investments that affect over 270 Army systems.

Prior to this assignment, Dr. Bochenek served as Deputy Program Executive Officer (DPEO) for Combat Support and Combat Service Support (CS&CSS). As DPEO, she led over 400 civilian personnel, 3 Project Management Offices, 18 managers, 250 ACAT III programs in five geographically distributed sites supporting the Army's tactical wheeled vehicle fleet and force projection commodities. Additionally in this role, Dr. Bochenek provided scientific and technical leadership, developed innovative programs for engineering career development, guided decisions on all milestone decision authority acquisition documents and advised civilian senior leadership on the development of the Army's Truck Modernization program, the Army's Long Term Armor Strategy,

[*]As of July 2010. Appendix includes bios distributed at the symposium.

and the Army's Future Tactical Truck Systems program. Dr. Bochenek's leadership also ensured the alignment of PEO goals with Army goals by developing and implementing the Strategic Readiness System Balanced Score Card metrics to meet organizational and programmatic goals. Prior to PEO CS&CSS, Dr. Bochenek was appointed to the Senior Executive Service as the Executive Director of Research and Technical Director for RDECOM-TARDEC. There, she led programs to align all ground-based systems science and technology research objectives to meet the Army's future warfighting and logistics needs including vehicle survivability, robotics, vehicle electronics, hybrid electric, alternative power and energy, and software engineering. In this dual-hatted role, she also was responsible for Science and Technology strategic planning, program selection, resource management, policy development, professional leadership and organizational liaison with the Director. She was also responsible for developing cooperative programs and agreements with industry, academia and other government agencies that facilitate exchange of technical intelligence.

RALPH BRODD

Ralph James Brodd is the new Director of the Kentucky-Argonne National Battery Manufacturing Research and Development Center. The center is a partnership among Argonne National Laboratory, the Commonwealth of Kentucky, the University of Louisville and the University of Kentucky. Brodd is also the President of Broddarp of Nevada, Inc., a consulting firm specializing in technology assessment, strategic planning and battery technology, production, and marketing. He received a B.A. degree in chemistry from Augustana College, Rock Island, Illinois, and M.A. and Ph.D. degrees in physical chemistry from the University of Texas at Austin.

Dr. Brodd began his career at the National Bureau of Standards in Washington, D.C., studying electrode reactions and phenomena that occur in battery operation. He taught physical chemistry in the U.S. Department of Agriculture Graduate School and lectured in electrochemistry at Georgetown University and American University. In the 1960s and 1970s, Dr. Brodd served in a variety of technical and management capacities with a number of battery companies. In 1961, Dr. Brodd joined the L.T.V. research Center of Ling Temco Vought, Inc., in Dallas, Texas, where he established a group in fuel cells and batteries. In 1963, he moved to the Battery Products Technology Center of Union Carbide Corporation, with technical management responsibilities for nickel-cadmium and lead acid rechargeable batteries, alkaline and carbon-zinc product lines, and exploratory R&D. He joined ESB (INCO Electroenergy, Inc.) in 1978, establishing a technology surveillance group, and moving to the position of Director of Technology with oversight and policy responsibility for R&D

laboratories serving product areas ranging from primary and secondary batteries to uninterruptible power supplies and small electric motors.

He was a member of the INCO Long Range Technology Committee and the technical advisory panel for North America Capital Venture Fund. In 1982, Dr. Brodd established Broddarp, Inc., a consulting firm specializing in battery technology, strategic planning, and technology planning. A consultancy with Amoco led to his moving to Amoco Research Center as project manager of a rechargeable lithium sulfur dioxide battery project. He subsequently moved to Gould, Inc., to establish their Lithium Powerdex Battery Venture and then to Valence Technology, a venture group developing a solid polymer electrolyte battery system for rechargeable batteries for portable consumer devices. He served as staff consultant/marketing director and then Vice President, Marketing.

Dr. Brodd was elected President of The Electrochemical Society in 1981 and Honorary Member in 1987. He was elected National Secretary of the International Society of Electrochemistry, 1977-1982, and Vice President, 1981-1983. He is past chairman of the Board of Directors of the International Battery Materials Association. Dr. Brodd was President of the Pi chapter of Phi Lambda Upsilon. Dr. Brodd has served on numerous technical advisory and review committees for the National Research Council, International Electrotechnic Commission, DOE, NASA, and NIH government laboratories and technical programs, most recently as a member of the 1999 and 2004 Review Committee for the Environmental Energy Technologies Division of Lawrence Berkeley National Laboratory. Dr. Brodd has over 100 publications and patents.

PATRICK DAVIS

Patrick Davis is the Program Manager of Energy Efficiency and Renewable Energy's Vehicle Technologies Program Office at the U. S. Department of Energy. The Vehicle Technologies Program supports over $200 million in annual research funding for hybrid drivetrains, advanced batteries, lightweight materials, advanced combustion and fuels, vehicle systems integration, and deployment activities. He is responsible for two major government industry partnerships, the FreedomCAR and Fuel Partnership and the 21st Century Truck Partnership. He also serves on the Board of Directors of the American National Standards Institute. Formerly he served as a senior advisor for transportation technologies in the office of Energy Efficiency and Renewable Energy and as DOE coordinator of the President's 20-in-10 Initiative to reduce gasoline usage in the United States by 20 percent in the next 10 years. He previously served as the Acting Program Manager of the Office of Hydrogen, Fuel Cells and Infrastructure Technologies, Team Leader for Hydrogen Production, Team Leader for Fuel Cell Technology, co-chair for two FreedomCAR and Fuel

Partnership Technical Teams, and the U.S. representative to the International Energy Agency's Hydrogen Implementing Agreement. Mr. Davis is a Chemical Engineer with 25 years of experience in the development of vehicle, alternative fuel, and electrochemical technologies.

JASON FORCIER

Jason Forcier has served as Vice President of our Automotive Solutions Group since August 2009. Prior to A123 Systems Mr. Forcier was named VP and GM of Lear's Global electronics business in August, 2008. Mr. Forcier served as regional president, automotive electronics for Robert Bosch LLC since April of 2007. In this position, Forcier was responsible for operations of Bosch's North American automotive electronics business.

He previously held the position of president and CEO of ETAS, Inc., a Bosch subsidiary. Mr. Forcier joined ETAS in 1997, where he was manager of engineering service. He has held various other assignments at ETAS, including a product line manager, manager of customer value teams and vice president of automotive embedded control tools. Prior to joining ETAS, Mr. Forcier worked in marketing at DSP Technologies and as a systems engineer at Delphi Automotive.

Mr. Forcier earned his bachelor's degree in mechanical engineering from the GMI Engineering & Management Institute (now Kettering University) and a master's degree in business administration from the University of Michigan. He has been a member of the Society of Automotive Engineers since 1990 and has participated in two previous panels at the SAE AVL Theater. Additionally, he has participated in A World in Motion, sponsored by the SAE Foundation.

LINDA GAINES

Linda Gaines is a Systems Analyst at the Center for Transportation Research at Argonne National Laboratory. She holds a B.A. in Chemistry and Physics from Harvard and a Ph.D. in Physics from Columbia. Her primary interest is in problem solving, applied to efficient use of resources. She began her 30+ years at Argonne by writing a series of handbooks of energy and material flows in the petroleum refining, organic chemicals, and copper industries that provided background for studies of technical and institutional issues involved in recycling discarded tires, packaging, and other energy-intensive materials. Dr. Gaines has examined the costs and impacts on energy use and the environment of production and recycling of advanced-design automobiles, trucks, and trains, and batteries. She has also examined the potential growth of electricity demand by industry and performed technical and economic analysis of alternative fuels, including hydrogen and biofuels. Her most recent work has involved studying ways to reduce petroleum use and other impacts from transport by recycling of batteries and also by reducing vehicle idling.

NANCY GIOIA

Nancy Lee Gioia is Ford Motor Company's Director of Global Electrification. Appointed to this position Oct. 9, 2009, Gioia directs strategy and planning for the next generation of Ford's global electric vehicle portfolio, touching all aspects of electrified transportation, including product planning, supplier partnerships and collaboration with the energy industry and government.

Prior to taking her current role, Gioia was Ford's director of Sustainable Mobility Technology and Hybrid Vehicle Programs for North America, overseeing research, development and ultimately deployment of other sustainable mobility technologies such as hydrogen internal combustion engines and hydrogen fuel cell vehicles.

Earlier in her career with Ford, Gioia was director of Current Model Vehicle Quality for North America where she was responsible for overall current model quality performance. Gioia has held several key management and executive positions within Ford Product Development, including valuable experience in electronics architecture design and integration in vehicles. She also was chief engineer for the Ford Aeromax class-8 truck line and the 2002 Ford Thunderbird program. She went on to direct engineering for all Ford, Lincoln and Mercury passenger cars in North America before taking on her current assignment.

Gioia joined Ford Motor Company in 1982 as a graduate trainee in the Electronics division. From 1983 to 1986, she held various positions in the division's Powertrain Business unit. In June 1986, she became manufacturing and quality engineer at the Engine Control Electronics facility in Lansdale, Pa. Her manufacturing experience continued in management positions in the organization, including the launch of Ford's new facility in Cadiz, Spain in 1989.

In June 1991, Gioia was named alliance manager for the Electronics division, responsible for the management, development and growth of more than 18 strategic alliances. In February 1993, she became Manufacturing and Materials Planning and Logistics manager, and, later that year, manager of assembly operations in the Climate Control division at Ford's Plymouth, Mich. plant.

Beginning in August 1994, Gioia took on a number of key assignments as engineering chief for several vehicle nameplates and platforms. She was named chief engineer – Commercial Truck, Automotive Components division and in

February 1996 was appointed chief program engineer for the Louisville/Aeromax truck line. She then served as chief program engineer for the all-new 2002 Ford Thunderbird, delivering the vehicle from concept to production.

Gioia combines her hands-on and management experience in electronics architectures, manufacturing, vehicle engineering, vehicle program management, quality engineering systems and executive direction to the Sustainable Mobility Technology and Hybrid Vehicle Programs group. This position includes direction of scientists working in Ford's Research and Innovations Center developing tomorrow's propulsion solutions and direction of a product engineering group applying and integrating new technologies into products for consumers today and in the future.

Gioia received her bachelor's degree in electrical engineering from the University of Michigan and her master of sciences in manufacturing systems engineering from Stanford University. While studying with the assistance of a Ford Advanced Education Fellowship, she received the Outstanding Service Award from the Stanford Institute for Manufacturing and Automation. In July 2001, she received the All Star Award from Automotive News and in 2005 she was named one of Automotive News' "100 Leading Women in the Auto Industry." She remains an active member of the Stanford University Alliance for Innovative Manufacturing (former chair) and is a board member of Auto Alliance International.

MARY GOOD

Dr. Mary L. Good, founding Dean and Donaghey Professor, is well known for her distinguished career. She has held many high-level positions in academia, industry, and government. The 143,000-member American Association for the Advancement of Science (AAAS) elected Dr. Good to serve as the president, following Dr. Stephen Jay Gould. In 2004, Dr. Good was recipient of the National Science Foundation's highest honor, the Vannevar Bush Award. She was also the first female winner of the AAAS's prestigious Philip Hogue Abelson prize for outstanding achievements in education, research and development management, and public service, spanning the academic, industrial, and government sectors. Two of her more than 27 awards include the National Science Foundation Distinguished Service medal and the esteemed American Chemical Society Priestly Medal. She is also the 6th Annual Heinz Award Winner. During the terms of Presidents Carter and Reagan, Dr. Good served on the National Science Board and chaired it from 1988-1991. She was the Undersecretary for Technology in the U.S. Department of Commerce and Technology during President Clinton's first term. This agency assists American industry to advance productivity, technology, and innovation in order to make

U.S. companies more competitive in the global market. Dr. Good has received 21 honorary degrees. Her undergraduate degree in chemistry is from the University of Central Arkansas. She earned her doctoral degree in inorganic chemistry from the University of Arkansas, Fayetteville, at age 24. Dr. Good spent 25 years teaching and researching at Louisiana State University and the University of New Orleans before becoming a guiding force in research and development for Allied Signal. Dr. Good was voted one of Arkansas' Top 100 Women by Arkansas Business.

WILLIAM HARRIS

Dr. Harris is the President and Chief Executive Officer of Science Foundation Arizona (SFAz). Prior to joining SFAz, Dr. William C. Harris was in Ireland serving as director general of Science Foundation Ireland (SFI), a new Irish agency that helped facilitate tremendous growth in Ireland's R&D sector during Harris' tenure. Immediately prior to going to Ireland, Dr. Harris was vice president of research and professor of chemistry and biochemistry at the University of South Carolina (USC). There, he oversaw research activities throughout the USC system, several interdisciplinary centers and institutes, the USC Research Foundation and sponsored research programs.

Dr. Harris served at the U.S. National Science Foundation (NSF) from 1978 to 1996, including as the director for mathematical and physical sciences (1991-1996). He was responsible for federal grants appropriation of $750 million. He also established 25 Science and Technology Centers to support investigative, interdisciplinary research by multi-university consortia. Earlier in his career, he catalyzed the Research Experience for Undergraduates program in the chemistry division and it became an NSF-wide activity.

In 2005, Dr. Harris was elected a member of the Irish Royal Academy, and received the Wiley Lifetime Achievement Award from California Polytechnic State University. He has authored more than 50 research papers and review articles in spectroscopy and is a fellow of the American Association for the Advancement of Science. Dr. Harris earned his undergraduate degree at the College of William and Mary, and received his Ph.D. in chemistry from the University of South Carolina.

DAVID HOWELL

Mr. Dave Howell is the Team Lead for the Hybrid Electric Systems Team at the Office of Vehicle Technologies Program, U.S. Department of Energy Headquarters, in Washington DC. Earlier, he was Manager, Electrochemical Energy Storage Research and Development, a position that he had been in since

2003. For the past 6 years he has also served as the DOE Co-Chair of the FreedomCAR Electrochemical Energy Storage Tech Team. Dave was a member of the research staff of the Oak Ridge National Laboratory (ORNL) in Oak Ridge, Tennessee for 12 years prior to joining DOE. At ORNL, he served as Project Manager for Aerospace Technologies. His primary focus was the development of advanced materials and processing techniques for aerospace structures. Dave served on active duty for 6 years at Wright Patterson AFB, Ohio. Dave was assigned as the Program Manager for Advanced Materials for Space Structures at the Air Force Materials Laboratory. In that role, he managed the Strategic Defense Initiative Organization's Advanced Materials for Space Structures Program supporting advanced materials R&D for spacecraft structures and mechanisms. Dave received a Bachelor of Science degree in Aerospace Engineering in 1985 from the University of Tennessee at Knoxville.

ROBERT KAMISCHKE

Robert Kamischke serves the Ener1 team as EnerDel's CFO, Controller, and CIO. His financial leadership and strategic planning acumen has played an instrumental role in EnerDel's rapid transformation from technology start-up firm to a full commercial volume producer of Lithium Ion Battery Energy Storage Systems.

Kamischke brings a breadth of experience from an award winning career where he led high performance teams for several leading automotive organizations. Most recently, Robert had a distinguished career at General Motors where his range of assignments were as diverse as leading all finance activities at the Pontiac, MI, full-size truck assembly plant to creating and implementing the GM Service Parts Accessory Distributor network. Robert notably served as Director of Finance and Strategy for the storied GM EV1 components business unit which developed the Electric Drive Train and Battery Energy Storage Systems in the mid 1990s.

Whether it is operations, manufacturing, treasury, information systems, venture development, sales, marketing or distribution; Robert has continuously been on the leading edge of innovation while seamlessly employing sound management principals and common sensibility to drive organization success.

Robert holds a Bachelor's degree in Accounting from Northern Michigan University and an MS in Manufacturing Management from Kettering University.

SRIDHAR KOTA

Dr. Sridhar Kota is a Professor of Mechanical Engineering at the University of Michigan-Ann Arbor where he has been involved in teaching and research in Design and Manufacturing area for 23 years. He is currently on leave from the U of M serving as the Assistant Director for Advanced Manufacturing at the White House Office of Science and Technology Policy. He has authored over 200 technical papers, holds 25 patents and served as an engineering consultant to numerous organizations in manufacturing, automotive, aerospace and MEMS fields. He is the recipient of the ASME Machine Design Award and the ASME Leonardo Da Vinci award. He is the founding President and CEO of FlexSys Inc. – a small business engaged in bio-inspired design of aircraft wings, wind turbine blades and automotive systems. Kota's research was featured in New York Times, Discovery Channel, Science News, Aviation Week, Popular Science, and other popular press.

In his current role at OSTP, Dr. Kota coordinates Federal advanced manufacturing R&D across agencies and addresses issues related to R&D funding gaps, manufacturing competitiveness, technology development and commercialization.

BOB KRUSE

Bob Kruse is the Founding Principal of EV Consulting LLC and the former Executive Director of global vehicle engineering for hybrids, electric vehicles and batteries at General Motors. As part of GM's commitment to fuel diversity, Kruse's division addresses strategic national interests and climate change risk by developing innovative vehicle design, reshaping the workforce and forging partnerships both inside and outside the automotive industry. He and his team have played a key role in developing the Volt, an electric hybrid vehicle being developed by GM.

Kruse holds a bachelor's degree from Missouri University of Science and Technology and a master's in management from MIT's Sloan School. He led the development of all parts and subsystems for vehicles and general assembly engineering, as well as global powertrain integration, where he was responsible for the first hybrid powertrain developed for full-size trucks.

Kruse went on to direct vehicle integration engineering, which created some the best automobiles in GM's history, such as the new Chevy Malibu, Buick Enclave, Cadillac CTS and the current generation of full-size trucks. He also led the performance division that executed award-winning vehicles during his tenure, including the second-generation Cadillac CTS-V, the Chevy Cobalt and HHR turbo SS.

Kruse's team has also developed the largest automotive advanced technology battery lab in the world in Warren, Mich., and helped establish the Advanced Battery Coalition of Drivetrains laboratory at the University of Michigan. Together, the two facilities partner in a wide range of work, including addressing critical workforce shortage issues. Under Kruse, GM also partners with the University of Michigan to offer master's-level online distance learning for engineers studying electrification technologies.

ANDY LEVIN

As a Department of Labor & Economic Growth Deputy Director and Chief Workforce Officer for the State of Michigan, Andy Levin oversees a number of bureaus, including: Workforce Programs, Career Education, Labor Market Information & Strategic Initiatives, Michigan Rehabilitation Services, the Michigan Commission for the Blind and Michigan Commission on Disability Concerns.

Levin has brought a wealth of experience in workplace and labor-management programs and policy to DLEG. At the national AFL-CIO, the U.S. Department of Labor, and the Presidential Commission on the Future of Worker-Management Relations, Levin helped create programs and lead innovative policy campaigns to improve economic security for working families and create business-labor partnerships. In addition, Levin's experience studying and working in Haiti, India, and China has brought additional strength to the Governor's efforts to attract business to Michigan from all over the world.

Levin started his career advocating for nursing home workers throughout Michigan, and he has extensive experience in the fast-growing health care industry, where the need to match workers to good jobs is acute. He is currently Of Counsel at the Southfield law firm of Klimist, McKnight, Sale, McClow, and Canzano, P.C. Levin is a long-time champion for diversifying Michigan's economy, greatly increasing production and use of renewable energy, and providing excellent education and health care for all.

Andy Levin is the son of U.S. Representative Sander Levin and the nephew of U.S. Senator Carl Levin. A Berkley native, he currently lives in Bloomfield Township with his wife, Mary Freeman, and their four young children. He earned a Bachelor's degree from Williams College, a Masters degree in Asian Languages and Cultures from the University of Michigan's Rackham Graduate School and a Law degree from Harvard University.

CARL LEVIN

Carl Levin is the senior US Senator from Michigan. In an editorial about Carl Levin, the Detroit News wrote, "He has been above reproach personally and has stuck to his principles, even when they were unpopular. Principled leadership, no matter what political ideology it comes from, is sorely needed in Washington." TIME Magazine recently named Carl Levin one of "America 's 10 Best Senators," noting that "the Michigan Democrat has gained respect from both parties for his attention to detail and deep knowledge of policy."

Carl Levin has worked to strengthen Michigan's industrial economy. Levin proposed the American Manufacturing Initiative to ensure that our government aggressively fights for manufacturing in America so our manufacturers and workers can compete globally on a level playing field. American manufacturers are not competing against foreign companies; they are competing against foreign governments.

As a co-chair of the Senate Auto Caucus and the Senate Auto Parts Task Force, Levin has been one of the most insistent voices in Washington calling for strong action to open the world's markets to American goods. Levin has been a longtime advocate of programs that provide for joint government-industry partnerships in development of advanced vehicle technologies. These efforts led to the growth of the Army's National Automotive Center in Warren, Michigan, which has played an important role in developing advanced technologies for military use, often in conjunction with the private sector.

As co-chair of the Senate Great Lakes Task Force, Levin has fought to protect the environmental treasures of "the Great Lakes State," an irreplaceable natural resource for Michigan and the country. In 1990, Levin authored the Great Lakes Critical Programs Act to create new standards of environmental protection for Great Lakes waters. Levin also helped win passage of the Great Lakes Legacy Program in 2002 to clean up contaminated sediments, and he worked to secure funding to deal with foreign aquatic invasive species including zebra mussels, milfoil and Asian carp. A strong advocate for the creation of the Thunder Bay National Marine Sanctuary, Levin has obtained significant funding for it and introduced legislation in 2009 to expand the boundaries of the sanctuary to more than eight times its current size. The expansion would help preserve "Shipwreck Alley" for divers and historians, where dozens of ships sank in the waters of Lake Huron.

Carl Levin is the Chairman of the Senate Armed Services Committee, where he has earned a reputation as a strong supporter of our national defense, a tireless advocate on behalf of our service men and women, and an effective fighter against wasteful government spending. Senator Levin has championed efforts to reduce the threats to our nation and the world from the proliferation of weapons of mass destruction and the threats posed by terrorism. He supports the efforts of the military services to transform their forces, technology, and tactics to meet these threats. He has been an active supporter of improving U.S. security by cooperative threat reduction, including arms control agreements that reduce weapons of mass destruction, and has fought for efforts designed to reduce the threat of proliferation of nuclear, chemical and biological weapons.

Senator Levin opposed the U.S. invasion of Iraq in 2003 and has authored several bipartisan proposals aimed at changing U.S. policy in Iraq. While Americans have differing opinions about our policy in Iraq, there is broad support of our brave men and women in uniform. Levin spearheaded the successful effort to pass the Dignified Treatment of Wounded Warriors Act, an historic reform to improve the way we provide medical care and ongoing support for our troops and our veterans, enacted in early 2008.

In 2007, Levin pushed to secure passage of the Acquisition Improvement and Accountability Act, the most far-reaching acquisition reform measure approved by Congress in more than a decade. The act requires, for the first time, that private security contractors working in a war zone must comply with Defense Department regulations and directives issued by our military commanders. The act also establishes a new acquisition workforce fund to hire the employees needed to manage defense contracts properly.

In 2009, Senator Levin secured passage of the Levin-McCain Weapon Systems Acquisition Reform Act, to fix major problems with the way the Department of Defense buys major weapons systems. The Act establishes a new, independent director of cost assessment to ensure that senior Pentagon managers have unbiased data to analyze project costs and cost projections. It also includes strengthening assessments of technologies that are under development and requiring the Department of Defense to conduct preliminary design reviews in advance of approving new acquisition programs.

The Franklin and Eleanor Roosevelt Institute awarded Senator Carl Levin its 2007 Four Freedoms Medal for his bipartisan efforts to reassert the role of the U.S. Senate in critical issues of foreign and military policy and for his longtime service to the country. The award recognizes Levin as "a leader dedicated to making government more effective, who holds himself and his colleagues to high ethical standards and insists that these same standards must apply to all facets of our society, both public and private; a leader whose efforts to strengthen America's armed forces have helped make the United States Military the finest fighting force in the world."

The National Guard Association of the United States presented Senator Levin with its 2004 Harry S. Truman Award for distinguished service in support of national defense. The award cited Levin's "long-standing, diligent and impassioned commitment on the readiness, morale and welfare of our military forces, their families and the modernization of our armed forces" that has had an "unparalleled and direct positive impact to the defense capabilities of the National Guard." In January 2003, the Secretary of the Navy cited Levin's "exceptional service to the Navy and Marine Corps" in presenting him its Distinguished Public Service Award, the highest award given to a civilian.

In July 2007, the President of Poland, Lech Kaczynski, presented Senator Levin with the Commander's Cross with the Star of the Order of Merit of the Republic of Poland. Instituted by Parliament in 1974, the award is conferred on foreigners and Polish residents abroad for service rendered to Poland.

As Chairman of the Permanent Subcommittee on Investigations, the premier investigating subcommittee in the Senate, Levin has focused on issues that impact the wallets of most Americans, including unfair credit card practices and sky-high oil and natural gas prices. Levin chaired numerous hearings delving into abusive credit card industry practices that help keep families mired in debt. The effort culminated in the 2009 enactment of the Credit Card Accountability, Responsibility and Disclosure Act or Credit CARD Act, which bans unfair practices by credit card companies including preventing credit card companies from retroactively raising interest rates on people who play by the rules, forcing banks to restore a lower interest rate for late-payers who make six months of on-time payments, and prohibiting the charging of interest on debt that is paid on time.

Another recent investigation found that excessive speculation in oil and natural gas markets resulted in higher prices for consumers. Levin introduced the "Close the Enron Loophole Act" to put a cop on the beat to police prices in U.S. energy markets that, due to Enron and others, are now largely unregulated. Levin's leadership enabled Senate passage of an amendment in late 2007 to close the Enron loophole and its enactment into law in May 2008.

In 2002, Levin led Congress' most in-depth examination into the collapse of Enron. His investigation exposed how Enron used deceptive accounting and tax transactions to report better financial results than the company actually experienced. The subcommittee's investigative work contributed to the accounting and corporate reforms enacted in the Sarbanes-Oxley Act in July 2002. In 2002, Levin began a three year investigation into the mass marketing of abusive tax shelters by KPMG and other professional firms, which was cited by The Washington Post as "a path-breaking inquiry . . . that served as a road map for prosecutors." Levin's bipartisan bill to end the use of tax havens will end some of the worst abuses of our tax laws by companies and individuals who avoid paying their U.S. taxes by using places such as the Cayman Islands to create sham transactions and shell corporations.

Under Levin's leadership, the Permanent Subcommittee on Investigations has also conducted a comprehensive money laundering investigation, which led to the enactment of legislation to detect and stop money laundering and terrorist financing. Levin is also a member of the Small Business and Entrepreneurship Committee and an ex officio member of the Senate Select Committee on Intelligence.

Carl Levin believes we must expand educational opportunities for all Americans if our nation is to remain strong and productive. He has fought for increased funding for the Head Start preschool program, Title I for educationally disadvantaged students, and Pell Grants and loans for college and vocational school students. Senator Levin has been a strong advocate for the effective use of technology in K-12 schools and helped create the Consortium for Outstanding Achievement in Teaching with Technology, a groundbreaking Michigan partnership helping teachers master technology skills. He has been an enthusiastic supporter of School to Work programs, which have created a public-private partnership to prepare students for the demands of the modern workplace. He has won critical federal support for the Focus: HOPE Center for Advanced Technology, a world-class manufacturing training facility in Detroit.

Addiction to illegal drugs continues to plague our society. Senator Levin authored a provision in the Drug Abuse and Treatment Act of 2000 to enable qualified physicians to prescribe and dispense from their private offices - rather than centralized clinics – revolutionary, new anti-addiction medications such as buprenorphine that suppress the craving for heroin.

Carl Levin was born in 1934 in Detroit, where he graduated from Central High School. In 1956, he graduated with honors from Swarthmore College and graduated from Harvard University Law School in 1959. He practiced and taught law in Michigan until 1964 when he was appointed an assistant attorney general of Michigan and the first general counsel for the Michigan Civil Rights Commission. He then helped establish the Detroit Public Defender's Office and led the Appellate Division of that office, which has become the State Appellate Defender's Office.

He won election to the Detroit City Council in 1969, becoming its president in 1973 by winning the most votes citywide. In 1978, he won an upset victory over the number two Republican in the U.S. Senate. He was reelected in 1984, 1990, 1996, 2002 and 2008.

GREG MAIN

Greg Main is President and CEO of the Michigan Economic Development Corporation, the State of Michigan's lead economic development agency, and is responsible for executing Governor Granholm's economic development strategy. From 2003 through June 2008, Mr. Main served as President and CEO of i2E, Inc. of Oklahoma City, a leading technology commercialization program, with responsibility for directing efforts to assist start-up companies in attracting risk capital and securing private equity funding. From 1998 to 2002, he was a general partner with Chisholm Private Capital Partners, a $66 million venture capital firm in Oklahoma City and beginning in 1994, a partner in Intersouth Partners of Research Triangle, N.C. Appointed as Oklahoma Secretary of Commerce in 1991 after a national search to lead the state's recovery from the oil bust, Mr. Main served as the state's chief economic development officer, administering a $90 million budget encompassing 180 employees. He designed and implemented initiatives including the award-winning Oklahoma Quality Jobs program and Quality Jobs Investment Act. In addition, he was instrumental in establishing the Alliance for Manufacturing and the launch of the Oklahoma Capital Investment Board Venture Investing program.

Mr. Main, a Michigan native, began his economic development career as executive director and chief planner for the six-county, nonprofit Central Upper Peninsula Planning and Development Regional Commission (CUPPAD) in Escanaba for 13 years beginning in 1970. He joined the Michigan Department of Commerce as director of the Upper Peninsula office in 1983. From 1985 to 1990, his duties as director of the manufacturing development group included responsibility for marketing Michigan as a location for manufacturing investment. In that capacity, he directed State of Michigan offices in Brussels, Tokyo, Toronto and Lagos, Nigeria. He was deputy director of economic development in 1991 when he relocated to Oklahoma. Mr. Main was born in Belding, Michigan and grew up in Lansing. He graduated summa cum laude from Michigan State University in 1970 with a degree in urban planning. He has extensive training and post-graduate studies in general management, marketing management, business and real estate finance, sales and quality management. He is past chairman and president of the Oklahoma Venture Forum; Science Museum Oklahoma board member; Oklahoma Academy executive committee member; and Creative Oklahoma board member.

JOHN PELLEGRINO

Dr. John M. Pellegrino is the Director of the Sensors & Electron Devices Directorate (SEDD) of the Army Research Laboratory (ARL). Dr. Pellegrino holds a bachelor's degree in Physics from Gordon College, Boston, MA and Master's and a Doctoral degree in Physics from the University of Wisconsin Madison.

Prior to his appointment in September 1998 to the Director, SEDD, Dr. Pellegrino was the Chief, Electro-Optics and Acoustics Division and Associate Director for Sensors Research. He also served as Chief, ARL Signal and Information Processing Division, and Chief, Optical Processing Branch, Harry

Diamond Laboratories. He began his professional career as a Physicist in September 1981 at the Harry Diamond Laboratories.

Dr. Pellegrino serves as the Chair of the RDECOM Sensors and Power & Energy Technology Focus Teams. He regularly serves by invitation as conference chair, technical consultant for various programs, and as a member of various advisory boards and committees. These include serving as a member of the SPIE Board of Directors, Chair of the SPIE Symposia Committee, Chair of the Office of Secretary of Defense Energy and Power Technologies Initiative, Army member of the Defense Department Advisory Group on Electron Devices, and conferences and studies on sensors and sensor networking.

Dr. Pellegrino is a fellow of the International Optical Engineering Society (SPIE), and a Senior Member of the IEEE; he also a member of AAAS, Sigma Xi, and the Optical Society of America. He is recipient of the 2009 Meritorious Presidential Rank Award; twice recipient of the U.S. Army Research and Development Achievement Award (1994 & 1997), and a recipient of the Harry Diamond Laboratories Hinman Award for Technical Achievement (1986). He has authored and co-authored more than two dozen technical papers and reports, and is co-editor of the book Acousto-Optic Signal Processing.

MICHAEL REED

Michael "Mike" Reed joined Magna in April, 2009 as General Manager – Battery Divisions with responsibility for the start-up of Magna's North American lithium ion battery cell and pack manufacturing. He has over 40 years experience in the battery industry in various technical, operational and general management roles. His international experience includes transitioning advanced technology products and manufacturing from research and development to commercialization.

Prior to joining Magna, Mike served as President, Chief Executive Officer and member of the Board of Directors of Electro Energy Inc. an advanced battery technology company serving military and aerospace markets with significant R&D funding from the US Department of Defense and Department of Energy.

Earlier at EaglePicher, he served as Chief Operating Officer of EaglePicher Horizon Batteries, LLC, where he directed the start-up of an advanced technology battery facility in Beijing, China. At Johnson Controls, Mike served as Director of Engineering for the Battery Group and then expatriate General Manager of South American Battery Operations headquartered in Sao Paulo, Brazil. Before that, Mike held several technical, operations and general management positions at Exide Corporation and Delco Remy Division of General Motors Corporation.

Mike holds a BSChE in chemical engineering from Purdue University and a MBA from Indiana University.

ANN MARIE SASTRY

With expertise in numerical simulation and advanced materials characterization and design, Dr. Ann Marie Sastry's teams work in applied energy technologies, and on fundamental problems in applied mathematics, biology, and electrochemistry. In education and workforce issues, Sastry has led development of novel curricula to address critical national energy needs.

Sastry and her collaborators have published over 60 peer-reviewed journal articles and book chapters, and she has delivered over 50 invited seminars at academic institutions and organizations, including the National Academy of Sciences and the National Institute of Health. Her work has featured in Nature, Business Week, and other publications. In energy technologies, her laboratory has developed new materials, invented techniques for the manufacture and optimization of batteries, and algorithms for optimization of power systems. Her laboratory's projects, sponsored by General Motors, the DoE, the Army Research Office, the Air Force Office of Scientific Research, NSF, the Keck Foundation, and the Ford Motor Company, include numerical simulation of performance of Li batteries for electric vehicles, design of microbatteries for implantable systems, creation of biological batteries comprised of cellular organelles coupled with engineered substrates, and modeling of fully integrated structural batteries for realization of multifunctional, composite materials.

Sastry's laboratory partners with university, national laboratory, and industrial workers to address problems of societal significance. These strategic partnerships include the GM/UM Advanced Battery coalition for Drivetrains, a center founded to speed technology insertion of storage technologies into electric vehicles, using advanced simulation, experimentation, optimization, and controls of batteries.

Sastry holds MS and PhD degrees from Cornell University, and a BS from the University of Delaware, all in Mechanical Engineering. She is the recipient of numerous honors for her work, including the 2007 ASME Gusts Larson Award, the University of Delaware Presidential Citation for Outstanding Achievement (2004), the UM college of Engineering 1938E (2000), the University of Michigan Harry Russel Award (1999), and NSF's Presidential Early Career Award for Scientists and engineers (1997). In 2005, she was honored with a University of Michigan Faculty Recognition Award, acknowledging outstanding contributions as a senior faculty member in research, teaching and service. She has served on three Editorial Boards: the ASME Journal of Engineering

Materials and Technologies, Journal of Composite Materials, and as a Founding Associate Editor of the Journal of the Mechanical Behavior of Biomedical Materials.

SUJAI SHIVAKUMAR

Dr. Sujai Shivakumar is a Senior Program Officer at the National Academies Board on Science, Technology, and Economic Policy. He conducts a portfolio of research on national policies that advance science, technology, and innovation. This includes a review of high technology public-private partnerships in the United States, an analysis of the drivers of productivity growth in the New Economy, and an assessment of national innovation policies in both developed and emerging economies.

Before joining the National Academies, Dr. Shivakumar conducted post-doctoral research at Indiana University's Workshop in Political Theory and Policy Analysis. He participated in a major study led by Elinor Ostrom on the role of incentives in the structure of development aid. An expert in the field of Constitutional Political Economy, Dr. Shivakumar has also advised the governments of Somaliland and Nepal on institution building and the development of sustainable national constitutions.

Dr. Shivakumar is the author of The Constitution of Development, Crafting Capabilities for Self-Governance, published by Macmillan, and The Samaritans' Dilemma, The Political Economy of Development Aid, coauthored with Elinor Ostrom, Clark Gibson, and Krister Andersson, and published by Oxford University Press. He is also co-editor with Charles Wessner of a National Research Council report on India's Changing Innovation System.

DANIEL SPERLING

Daniel Sperling is Professor of Civil Engineering and Environmental Science and Policy, and founding Director of the Institute of Transportation Studies at the University of California, Davis (ITS-Davis). The Institute is staffed by over 150 faculty, staff, and student researchers. He is also Acting Director of the UC Davis Energy Efficiency Center. In February 2007, Governor Schwarzenegger appointed Dr. Sperling to the "automotive engineering" seat on the California Air Resources Board. His appointment was confirmed by the California Senate in January 2008. His chief responsibilities are oversight and design of the state's climate change, alternatives fuels, vehicle travel and land use, and zero emission vehicle programs. He also served as co-director of the California Low Carbon Fuel Standard study, requested in the Governor's January 2007 Executive Order.

In 2008 he was appointed chair of the "Future of Mobility" Council of the Davos World Economic Forum.

Dr. Sperling has led ITS-Davis to international prominence by building strong partnerships with industry, government, and the environmental community, integrating interdisciplinary research and education programs, and connecting research with public outreach and education. ITS-Davis won the 2006 Robert M. Zweig Public Education Award of the National Hydrogen Association, 2005 TRANNY award for Organization of the Year by the California Transportation Foundation, 1998 Employer of the Year Award of the Women's Transportation Seminar of Sacramento, and was selected as a finalist for the 2003 World Technology Energy Award. Dr. Sperling is recognized as a leading international expert on transportation technology assessment, energy and environmental aspects of transportation, and transportation policy. He has testified ten times to the US Congress and state legislatures, and provided keynote presentations and invited talks in recent years at international conferences in Asia, Europe, and North America. In the past 25 years, he has authored or co-authored over 200 technical papers and 11 books, including *Two Billion Cars* (Oxford University Press, 2009).

He was "lead author" of the transportation chapter in the 2007 IPCC report, "Mitigation of Climate Change," (IPCC won the Nobel Peace Prize in 2008) and a recent member of 13 National Academies committees on Energy Efficiency, Gasoline Taxes, Hydrogen, Transport in China, Biomass Fuels R&D, Sustainable Transportation, and related topics. He was founding chair of standing committees for the U.S. Transportation Research Board on Alternative Transportation Fuels (1989-'96), and Sustainability and Transportation (2006-08). He is the founding organizer of the premier conference on transportation and energy policy, bringing together every two years since 1988 the leaders from industry, government, academia, and the environmental community. He serves on many advisory committees and advises senior executives of many automotive and energy companies, environmental groups, and national governments, including review committees at three DOE national laboratories. He is widely cited in leading newspapers, has been interviewed many times on NPR radio, including *Science Friday*, *Talk of the Nation*, and *Fresh Air*, and in February 2009 he was featured on *The Daily Show with Jon Stewart*.

He received the following awards: 2009 Robert Zweig Public Education Award of the National Hydrogen Association, National Associate member of the National Academies in 2004, 2002 Carl Moyer Memorial Award for Scientific Leadership and Technical Excellence by the Coalition for Clean Air, 1997 "Clean Air Award" by the American Lung Association of Sacramento, 1996 Distinguished Public Service Award by the University of California, Davis, and 1993 Gilbert F. White Fellowship by Resources for the Future (Washington, D.C.). Prior to obtaining his Ph.D. in Transportation Engineering from the

University of California, Berkeley (with minors in Economics and Energy & Resources), Professor Sperling worked two years as an environmental planner for the US Environmental Protection Agency and two years as an urban planner in the Peace Corps in Honduras. He has an undergraduate degree in engineering and urban planning from Cornell University. During 1999-2000, he was on leave as a visiting scholar at OECD (European Conference of Ministers of Transport).

DEBBIE STABENOW

Born and raised in Michigan, United States Senator Debbie Stabenow knows what matters to Michigan. She made history in 2000 when she became the first woman from the State of Michigan elected to the United States Senate. From the County Commission to the State Legislature to the halls of Congress, she is a respected national leader on health care and manufacturing issues and champion for Michigan. She has risen in Senate leadership as Senate Conference Secretary and now Chair of the Democratic Steering and Outreach Committee.

A nationally recognized leader, Senator Stabenow is respected for her ability to build coalitions to get things done for Michigan and our nation. Her recent appointment to the Senate Energy and Natural Resources Committee, and membership on the Senate Finance, Agriculture and Budget Committees, has given her a powerful and unique role to play in shaping our nation's health care, manufacturing and energy policies, so critical to our future. Senator Stabenow is fighting for new laws to crack down on countries violating our trade laws and to reduce health care costs. She is a recognized leader in the fight to make prescription drugs more affordable and to bring innovative technology to the health care system. Her proposed Green Collar Jobs Initiative would retool older manufacturing facilities and invest in the newest energy technologies, including advanced batteries, to reduce our dependence on foreign oil and create jobs here at home.

She is also delivering for Michigan as a member of the Senate Agriculture Committee. Her leadership in rewriting our nation's farm bill has brought about an historic new focus on Michigan's specialty crops and victories for Michigan in alternative energy production, Great Lakes preservation, land conservation, research, food safety, nutrition, and rural development. Senator Stabenow also wrote a new law to assist families facing foreclosure by eliminating the IRS rule that unfairly taxed homeowners who had a portion of their original mortgage loan forgiven by the bank. She is a strong advocate for higher education – working to pass recent increases in college financial aid and securing millions in cutting-edge research dollars for Michigan's colleges and universities. She is a champion for Michigan's Great Lakes and waterways, and author of the Michigan Lighthouse and Maritime Heritage Act to promote tourism and help preserve some of Michigan's greatest historical treasures. She is also the author of the first ever federal ban on drilling for oil and gas in our Great Lakes.

As a State Legislator, Stabenow was acclaimed one of Michigan's most passionate advocates for children and an expert in family law and small business issues. Her influence as a State Legislator is evident throughout Michigan law – from Michigan's historic property tax cut and small business reforms, to nationally acclaimed legislation to protect children and families.

DAVID STIEREN

David Stieren is the Manager of Technology Acceleration for the Manufacturing Extension Partnership (MEP) at the National Institute of Standards and Technology (NIST). Mr. Stieren is responsible for the development and deployment of the processes used by the National MEP System to accelerate the many different ways by which U.S. manufacturers can leverage technology to their competitive advantage. MEP operates 59 Centers and over 440 service locations in all 50 U.S. states and Puerto Rico, providing assistance to approximately 30,000 U.S. manufacturers on an annual basis.

Mr. Stieren has extensive knowledge of the operations, systems, and technologies used by a broad array of U.S. manufacturing industries, from defense and aerospace, to automotive, shipbuilding, semiconductor electronics, heavy equipment, fuel cell, and many others. He has significant experience developing and managing strategic and technical partnerships involving U.S. industry, government agencies, and academia. Mr. Stieren has a Bachelor's Degree in Mechanical Engineering, a Master's Degree in Technology Management, and he completed the coursework for a Doctorate Degree in Engineering Management.

BILL VAN AMBURG

Bill Van Amburg is Senior Vice President for WestStart-CALSTART, a non-profit, fuel-neutral and member-supported consortium of more than 145 companies worldwide. It focuses on helping companies and agencies develop and produce clean and efficient vehicles, advanced and renewable fuels and new systems for transit and personal mobility, serving as a strategic broker to the industry.

Bill oversees teams in five program areas at the consortium: Heavy hybrids; New Fuels; Technology commercialization; Fleet analysis and consulting; and industry services. Responsibilities include the Hybrid Truck Users Forum (HTUF), a national program to speed the production and commercialization of heavy-duty hybrid trucks, operated in a partnership with the U.S. Army's National Automotive Center (NAC), with support from the Hewlett Foundation and the Department of Energy (DOE). Hybrid trucks and buses reduce fuel use and emissions, saving users money while contributing to reductions in foreign oil importation and global warming emissions. He is also involved with WestStart-CALSTART projects focused on overall greenhouse gas reduction and energy security strategies.

Van Amburg brings more than 25 years of experience in marketing and market development, technology commercialization, communications and environmental markets, including emission credit trading. Previously, Van Amburg was senior vice president with the first electronic emission credit exchange, has operated his own environmental marketing consulting practice, The Ardent Group, as well as serving previously as a vice president with WestStart-CALSTART from 1993-2000. Prior to that he had a nearly two decade career as an Emmy award winning broadcast journalist focusing on science, technology and environmental issues.

He is a graduate of the Executive Management Program at the UCLA Anderson School of Management and has a certificate in Brand Management from the Stanford Alumni Association, as well as a bachelor's degree in Anthropology from the University of California, Berkeley.

CHARLES WESSNER

Charles Wessner is a National Academy Scholar and Director of the Program on Technology, Innovation, and Entrepreneurship. He is recognized nationally and internationally for his expertise on innovation policy, including public-private partnerships, entrepreneurship, early-stage financing for new firms, and the special needs and benefits of high-technology industry. He testifies to the U.S. Congress and major national commissions, advises agencies of the U.S. government and international organizations, and lectures at major universities in the U. S. and abroad. Reflecting the strong global interest in innovation, he is frequently asked to address issues of shared policy interest with foreign governments, universities, research institutes, and international organizations, often briefing government ministers and senior officials. He has a strong commitment to international cooperation, reflected in his work with a wide variety of countries around the world.

Dr. Wessner's work addresses the linkages between science-based economic growth, entrepreneurship, new technology development, university-industry clusters, regional development, small-firm finance and public-private partnerships. His program at the National Academies also addresses policy issues associated with international technology cooperation, investment, and trade in high-technology industries.

Currently, he directs a series of studies centered on government measures to encourage entrepreneurship and support the development of new technologies and the cooperation between industry, universities, laboratories, and government to capitalize on a nation's investment in research. Foremost among these is a congressionally mandated study of the Small Business Innovation Research (SBIR) Program, reviewing the operation and achievements of this $2.3 billion award program for small companies and start-ups. He is also directing a major study on best practice in global innovation programs, entitled *Comparative Innovation Policy: Best Practice for the 21st Century.* Today's meeting on "Clustering for 21st Century Prosperity" forms part of a complementary analysis entitled *Competing in the 21st Century: Best Practice in State & Regional Innovation Initiatives*. The overarching goal of Dr. Wessner's work is to develop a better understanding of how we can bring new technologies forward to address global challenges in health, climate, energy, water, infrastructure, and security.

SONYA ZANARDELLI

Sonya Zanardelli is the US Army RDECOM-TARDEC Ground Vehicle Power & Mobility, Energy Storage Team Leader & DOD Power Sources Member. Sonya Zanardelli received her B.S. from Wayne State University and M.S. degree in Electrical Engineering from University of Michigan - Dearborn, in 2002 and 2005, respectively. She is currently working at US Army Tank Automotive Research Development Engineering Center (TARDEC) in Warren, MI and holds the position of Energy Storage Team Leader in the Research Business Group under the Ground Vehicle Power & Mobility Directorate and has worked at TARDEC for 9.5 years. Her research fields of interest include bidirectional converters and control and advanced energy storage research for military ground vehicle applications.

Appendix C

Participants List

Mohamed Alamgir
Compact Power

Giedrius Ambrozaitis
Alliance of Automobile Manufacturers

Dave Andrea
Original Equipment Suppliers Association

Robert Bachrach
Applied Materials

Gretchen Baier
The Dow Chemical Company

Jay Baron
Center for Automotive Research

Daniel Beattie
Clark Hill

Jeff Benton
Quallion LLC

Earl Bloom
Dow Kokam

Jeff Bocan
Beringea, LLC

Grace Bochenek
TARDEC

Raymond Boeman
Oak Ridge National Laboratory

Gerri Botte
Ohio University

Ralph Brodd
Kentucky-Argonne National Battery Manufacturing R&D Center

Brad Brodie
DENSO International America

Michael Brundage
CERDEC

Satish Chikkanavar
University of Michigan

McAlister Clabaugh
National Academy of Sciences

Steve Clark
Chrysler Group, LLC

Stephen Creager
Clemson University

Claus Daniel
Oak Ridge National Laboratory

Patrick Davis
US Department of Energy

David Dawson
National Academy of Sciences

David Dierksheide
National Academy of Sciences

Lawrence Drzal
Michigan State University

Lindsey Eister
MEDC

Pete Engardio
National Academy of Sciences

Patrick Ennis
MEDC

Jason Forcier
A123 Systems

Linda Gaines
Argonne National Laboratory

Joshua Gardner
Edelman

Kevin Geiss
US Army

Nancy Gioia
Ford Motor Company

Mary Good
University of Arkansas at Little Rock

Joseph Gordon
Applied Materials

Gale Govaere
Office of Senator Levin

James Greenberger
NAATBATT

Alan Greenshields
fortu Holding AG

Chuck Gulash
Toyota Technical Center

Korey Hall
Office of Senator Stabenow

John Harb
Brigham Young University

William Harris
Science Foundation Arizona

Oliver Hazimeh
PRTM

Kwangtaek Hong
DENSO International America, Inc.

Greg Hopton
AVL

Jim Hotary
Faurecia Westworks

David Howell
US Department of Energy

Louis Infante
Ricardo Inc

Neil Johnson
Ricardo Inc

Erik Kallio
TARDEC

Robert Kamischke
ENERDEL, Inc

Kyle Kimel
AVL

Jerry Klarr
AVL

Kathy Kleckner
MEDC

John Koch
ECD/Ovonic

Sridhar Kota
White House Office of Science and Technology Policy

Steve Kraemer
FEV

Gary Krause
MEDC

Robert Kruse
General Motors

Johannes Kuhn
Detroit Diesel Corporation

Jeff LeBrun
Sakti3

Andy Levin
Michigan DLEG

Carl Levin
US Senate

Dana Lowell
Faurecia Westworks

Greg Main
MEDC

Jairam Manjunathaiah
MAG Renewable Energy Team

Don Manvel
AVL
Jeff Mason
University Research Corridor

Barb McCallahan
Office of Senator Stabenow

Roger Meister
fortu Holding AG

Bob Metzger
MEDC

James Miller
Argonne National Laboratory

Pravansu Mohanty
University of Michigan

Om Nalamasu
Applied Materials

Simon Ng
Wayne State University

Vince Nystrom
University Research Corridor

Bill Ott
Actacell

John Pellegrino
TARDEC

Bill Pine

Michael Psarouthakis
MEDC

Bob Purcell
Purcell and Associates

Saqib Rahim
ClimateWire

Chuck Reardon
Dow Kokam

Michael Reed
Magna

Kevin Rzemien
FEV

Jim Saber
NextEnergy

Jeffrey Sakamoto
University of Michigan

Ann Marie Sastry
University of Michigan

Dan Schmidt
Dow Kokam

Vicki Selva
Office of Senator Levin

Sujai Shivakumar
National Academy of Sciences

Michael Shore
MEDC

Eric Shreffler
MEDC

Paul Skalny
TARDEC

Tim Slusser
MEDC

Brett Smith
Center for Automotive Research

Gary Smyth
General Motors

Beth Sommers
State of Michigan

Dan Sperling
UC Davis

Debbie Stabenow
US Senate

David Stieren
NIST

Sandy Stojkovski
AVL

Joe Thompson
Nissan

Mohit Uberoi
MEGTEC Systems

Bill Van Amburg
CALSTART

Chao-Yang Wang
The Pennsylvania State University

Tom Watson
Johnson Controls

Roger Wery
PRTM

Charles Wessner
National Academy of Sciences

Kathleen White
MEDC

Joachim Wolschendorf
FEV

Kohei Yamaguchi
DENSO International America, Inc.

Sonya Zanardelli
TARDEC

Appendix D

Bibliography

Accelerate Arkansas Strategic Planning Committee. 2007. *Building a Knowledge-Based Economy in Arkansas: Strategic Recommendations by Accelerate Arkansas*. Teresa A. McLendon, ed. Little Rock: Accelerate Arkansas.

Acs, Z., and D. Audretsch. 1990. *Innovation and Small Firms*. Cambridge, MA: The MIT Press.

Alic, J. A., L. M. Branscomb, H. Brooks, A. B. Carter, and G. L. Epstein. 1992. *Beyond Spin-off: Military and Commercial Technologies in a Changing World*. Boston: Harvard Business School Press.

American Wind Energy Association. 2011. "Arkansas is a National Leader in Wind Energy Manufacturing." Washington, DC: American Wind Energy Association. August.

Amsden, A. H. 2001. *The Rise of "the Rest": Challenges to the West from Late-industrializing Economies*. Oxford: Oxford University Press.

Arizona Daily Star. 2009. "Budget Cuts Hit Science Research Partnerships at Arizona Universities." February 8.

Arkansas Business. 1996. "Nucor Makes Blytheville Steel Capital of the South." December 16.

Arkansas Business. 2011. "Beebe, FDA Sign First of its Kind Agreement at NCTR." August 12.

Arkansas Business. 2011. "NCTR Has Potential to Create High-Paying Jobs." July 4.

Arkansas Business. 2012. "AMS Grant Helps Local Aerospace Manufacturer Turn Business Around." January 5.

Arkansas Democrat-Gazette. 1987. "Arkansas Legislators Present Their Proposal for Tax Breaks for Proposed Steel Mill." December 7.

Arkansas Democrat-Gazette. 1999. "Results from Subsidies Unknown–State Has Little Idea Whether $633 Million in Breaks to Firms Spurred Investment." December 12.

Arkansas Department of Education. 2007. *Combined Research Report of Business Leaders and College Professors on Preparedness of High School Graduates.* January. Little Rock: Arkansas Department of Education.

Arkansas Economic Development Commission. 1979. *Arkansas Climbs the Ladder: A View of Economic Factors Relating to Growth of Jobs and Purchasing Power*. Little Rock: Arkansas Economic Development Commission.

Arkansas Economic Development Commission. 2002. "Report of the Task Force for the Creation of Knowledge-Based Jobs." Little Rock: Arkansas Economic Development Commission.

Arkansas Economic Development Commission. 2009. *Governor Mike Beebe's Strategic Plan for Economic Development.* Little Rock: Arkansas Economic Development Commission.

Arkansas Research and Education Optical Network. 2008. "Arkansas Cyberinfrastructure Strategic Plan." Little Rock. *<http://areon.net/resources/CyberinfrastructureStrategicPlan20081024.pdf>*

Arkansas Small Business and Technology Development Center. 2009. "Enterprise Center to Offer Valuable Technology Incubator Resources." Press Release. Little Rock: University of Arkansas at Little Rock.

Arkansas State University. 2011. "Brian Rogers Named Director of Commercial Innovation Technology Incubator." Press Release. Jonesboro, AR: Arkansas State University. January 5.

Arkansas Task Force on Higher Education Remediation, Retention and Graduation Rates. 2008. *Access to Success: Increasing Arkansas' College Graduates Promotes Economic Development.* ("Education Task Force Report.") August.

Arkansas Task Force on Higher Education Remediation, Retention, and Graduation Rates. 2008. *A Plan for Increasing the Number of Arkansans with Bachelor's Degrees*. Little Rock: Arkansas State University.

ArkansasOnline. 2008. "LM Glasifiber Dedicates Little Rock Factory." October 28.

Asheim, B., A. Isaksen, C. Nauwelaers, and F. Todtling, eds. 2003. *Regional Innovation Policy for Small-Medium Enterprises*. Cheltenham, UK, and Northampton, MA: Edward Elgar.

Athreye, S. 2000. "Technology policy and innovation: The role of competition between firms." In P. Conceicao, S. Shariq, and M. Heitor, eds. *Science, Technology, and Innovation Policy: Opportunities and Challenges for the Knowledge Economy*. Westport, CT, and London: Quorum Books.

Atkinson, R., and S. Andes. 2010. *The 2010 State New Economy Index: Benchmarking Economic Transformation in the States.* Washington, DC: Kauffman Foundation and the Information Technology and Innovation Foundation. November.

Audretsch, D., ed. 1998. *Industrial Policy and Competitive Advantage, Volumes 1 and 2*. Cheltenham, UK: Edward Elgar.

Audretsch, D. 2006. *The Entrepreneurial Society*. Oxford: Oxford University Press.

Audretsch, D., B. Bozeman, K. L. Combs, M. Feldman, A. Link, D. Siegel, P. Stephan, G. Tassey, and C. Wessner. 2002. "The economics of science and technology." *Journal of Technology Transfer* 27:155-203.

Audretsch, D., H. Grimm, and C. W. Wessner. 2005. *Local Heroes in the Global Village: Globalization and the New Entrepreneurship Policies.* New York: Springer.

Augustine, C., et al. 2009. *Redefining What's Possible for Clean Energy by 2020.* Full Report. Gigaton Throwdown. June.

Bajaj, V. 2009. "India to spend $900 million on solar." *The New York Times* November 20.

Baldwin, J. R., and P. Hanel. 2003. *Innovation and Knowledge Creation in an Open Economy: Canadian Industry and International Implications.* Cambridge: Cambridge University Press.

Balzat, M., and A. Pyka. 2006. "Mapping national innovation systems in the OECD area." *International Journal of Technology and Globalisation* 2(1-2):158-176.

Bank of Boston. 1997. "MIT: The Impact of Innovation." Boston: Bank of Boston.

Battelle. 2009. *R&D Magazine*. December.

Battelle Technology Partnership Practice. 2009. "Opportunities for Advancing Job-Creating Research in Arkansas: A Strategic Assessment of Arkansas University and Government Lab Research Base." Access at *<http://www.aralliance.org/__data/assets/pdf_file/0017/1682/Job-Creating-Research-in-Arkansas.pdf>*.

Bezdek, R. H., and F. T. Sparrow. 1981. "Solar subsidies and economic efficiency." *Energy Policy* 9(4):289-300.

Biegelbauer, P. S., and S. Borras, eds. 2003. *Innovation Policies in Europe and the U.S.: The New Agenda.* Aldershot, UK: Ashgate.

Birch, D. 1981. "Who creates jobs?" *The Public Interest* 65:3-14.

Biris, Alexandru S. et al. 2009. "*In vivo* Raman flow cytometry for real-time detection of carbon nanotube kinetics in lymph, blood, and tissues." *J. Biomed Opt* 14(2).

Block, F., and M. R. Keller. 2008. *Where Do Innovations Come From? Transformations in the U.S. National Innovation System, 1970-2006.* Washington, DC: The Informational Technology & Innovation Foundation. July.

Blomström, M., A. Kokko, and F. Sjöholm. 2002. "Growth & Innovation Policies for a Knowledge Economy: Experiences from Finland, Sweden, & Singapore." EIJS Working Paper. Series No. 156.

Bloomberg News. 2006. "The next green revolution." August 21.

Bolinger, M., R. Wiser, and E. Ing. 2006. "Exploring the Economic Value of EPAct 2005's PV Tax Credits." Lawrence Berkeley National Laboratory.

Borenstein, S. 2008. *The Market Value and Cost of Solar Photovoltaic Electricity Production.* Berkeley, CA: Center for the Study of Energy Markets.

Borras, S. 2003. *The Innovation Policy of the European Union: From Government to Governance.* Cheltenham, UK: Edward Elgar.

Borrus, M., and J. Stowsky. 2000. "Technology policy and economic growth." In C. Edquist and M. McKelvey, eds. *Systems of Innovation: Growth, Competitiveness and Employment, Vol. 2*. Cheltenham, UK and Northampton, MA: Edward Elgar.

Bradsher, K. 2009. "China builds high wall to guard energy industry." *International Herald Tribune* July 13.

Brander, J. A., and B. J. Spencer. 1983. "International R&D rivalry and industrial strategy." *Review of Economic Studies* 50:707-722.

Brander, J. A., and B. J. Spencer. 1985. "Export strategies and international market share rivalry." *Journal of International Economics* 16:83-100.

Branigin, W. 2009. "Obama lays out clean-energy plans." *Washington Post* March 24. p. A05.

Branscomb, L., and P. Auerswald. 2002. *Between Invention and Innovation: An Analysis of Funding for Early-Stage Technology Development.* NIST GCR 02–841. Gaithersburg, MD: National Institute of Standards and Technology. November.

Braunerhjelm, P., and M. Feldman. 2006. *Cluster Genesis: Technology based Industrial Development.* Oxford: Oxford University Press.

Business Facilities. 2010. "Windstream Picks Little Rock, AR for HQ." July 13.

Bush, N. 2005. "Chinese competition policy, it takes more than a law." *China Business Review* May-June.

Bush, V. 1945. *Science: The Endless Frontier*. Washington, DC: Government Printing Office.

Bussey, J. 2012." The sun shines on 'the cloud.'" *The Wall Street Journal*. July 13:B1.

Campoccia, A., L. Dusonchet, E. Telaretti, and G. Zizzo. 2009. "Comparative analysis of different supporting measures for the production of electrical energy by solar PV and Wind systems: Four representative European cases." *Solar Energy* 83(3):287-297.

Caracostas, P., and U. Muldur. 2001. "The emergence of the new European Union research and innovation policy." In P. Laredo and P. Mustar, eds. *Research and Innovation Policies in the New Global Economy: An International Comparative Analysis*. Cheltenham, UK: Edward Elgar.

Carter, Mark. 2011. "Scholars Program Copies Georgia's Model." *Innovate Arkansas* August 22.

Chesbrough, H. 2003. *Open Innovation: The New Imperative for Creating and Profiting from Technology*. Cambridge, MA: Harvard Business School Press.

Chronicle of Higher Education. February 7, 2010.

Cimoli, M., and M. della Giusta. 2000. "The Nature of Technological Change and Its Main Implications on National and Local Systems of Innovation." IIASA Interim Report IR-98-029.

Clemson School of Computing. Undated. "Dr. Amy Apon Joins the School of Computing as Chair of the Computer Science Division." Press Release. Clemson, SC: Clemson University.

Cleveland.com. 2010. "Caterpillar Opens New Arkansas Factory, Hiring 600." September 1.

Coburn, C., and D. Berglund. 1995. *Partnerships: A Compendium of State and Federal Cooperative Programs*. Columbus, OH: Battelle Press.

Combs, K., and A. Link. 2003. "Innovation policy in search of an economic paradigm: the case of research partnerships in the United States." *Technology Analysis & Strategic Management* 15(2).

Computer Community Consortium. 2009. "From Internet to Robotics: A Roadmap for U.S. Robotics." May 21.

Constable, George, and Bob Somerville. 2003. *A Century of Innovatio: Twenty Engineering Achievements that Transformed our Lives.* Washington DC: Joseph Henry Press.

Cortright, J. 2006. *Making Sense of Clusters: Regional Competitiveness and Economic Development*. Washington, DC: Brookings Institution.

Cortright, J., and H. Mayer. 2002. *Signs of Life: The Growth of Biotechnology Centers in the US.* Washington, DC: Brookings Institution.

Council on Competitiveness/National Governor's Association. 2007. *Cluster-Based Strategies for Growing State Economies*. Washington, DC: Council on Competitiveness.

Crafts, N. F. R. 1995. "The golden age of economic growth in Western Europe, 1950-1973." *Economic History Review* 3:429-447.

Dahlman, C., and J. E. Aubert. 2001. *China and the Knowledge Economy: Seizing the 21st Century.* Washington, DC: World Bank.

Dahlman, C., and A. Utz. 2005. *India and the Knowledge Economy: Leveraging Strengths and Opportunities*. Washington, DC: World Bank.

Daniel, D. E. 2008. "Thoughts on Creating More Tier One Universities in Texas." White Paper. May 30.

Darmody, B. 2010. "The Power of Place 2.0: The Power of Innovation—10 Steps for Creating Jobs, Improving Technology Commercialization, and Building Communities of Innovation." Tucson: Association of University Research Parks. March 5.

Davis, S., J. Haltiwanger, and S. Schuh. 1993. "Small Business and Job Creation: Dissecting the Myth and Reassessing the Facts." Working Paper No. 4492. Cambridge, MA: National Bureau of Economic Research.

Debackere, K., and R. Veugelers. 2005. "The role of academic technology transfer organizations in improving industry science links." *Research Policy* 34(3):321-342.

Department of Labor and Industrial Relations: Research and Statistics Office. *Hawaii's Green Workforce: A Baseline Assessment.* December 2010.

Desai, S., P. Nijkamp, and R. R. Stough, eds. 2011. *New Directions in Regional Economic Development: The Role of Entrepreneurship Theory and Methods, Practice and Policy*. Northampton, MA: Edward Elgar.

De la Mothe, J., and G. Paquet. 1998. "National Innovation Systems, 'Real Economies' and Instituted Processes." *Small Business Economics* 11:101-111.

DeVol, R. C., K. Klowden, A. Bedorussian, and B. Yeo. 2009. *North America's High Tech Economy: The Geography of Knowledge-Based Institututions.* June 2.

De Vol, Ross et al. 2004. *Arkansas' Position in the Knowledge-Based Economy* Santa Monica: Milken Institute.

Dobesova, K., J. Apt, and L. Lave. 2005. "Are renewable portfolio standards cost-effective emissions abatement policy?" *Environmental Science and Technology* 39:8578-8583.

Doloreux, D. 2004. "Regional innovation systems in Canada: a comparative study." *Regional Studies* 38(5):479-492.

Doris, E., J. McLaren, V. Healey, and S. Hockett. 2009. *State of the States 2009: Renewable Energy Development and the Role of Policy*. Golden, CO: National Renewable Energy Laboratory.

Durham, C. A., B. G. Colby, and M. Longstreth. 1988. "The impact of state tax credits and energy prices on adoption of solar energy systems." *Land Economics* 64(4):347-355.

Eaton, J., E. Gutierrez, and S. Kortum. 1998. "European Technology Policy." NBER Working Paper 6827.

Economic Development Agency. 2011. "EDA Announces Registry to Connect Industry Clusters Across the U.S.; Harvard Business School Tool Designed to Assist Innovators and Small Business in Spurring Regional Economic Growth." October 6.

Edler, J., and S. Kuhlmann. 2005. "Towards one system? The European Research Area initiative, the integration of research systems and the changing leeway of national policies." *Technikfolgenabschätzung: Theorie und Praxis* 1(4):59-68.

Eickelpasch, A., and M. Fritsch. 2005. "Contests for cooperation: a new approach in German innovation policy." *Research Policy* 34:1269-1282.

Energy Information Administration. 2008. *Federal Financial Interventions and Subsidies in Energy Markets 2007*. Washington, DC: Energy Information Administration.

Energy Overviews. 2011. "Arkansas Wins $100 Million Wind Turbine Nacelle Plant." May 11.

Engardio, P. 2008. "Los Alamos and Sandia: R&D Treasures." *BusinessWeek*. February 11.

Engardio, P. 2009. "State Capitalism." *BusinessWeek*. February 6.

Etzowitz, H. 2008. *The Triple Helix: University-Industry-Government Innovation in Action*. London: Routledge.

European Commission. 2003. *Innovation in Candidate Countries: Strengthening Industrial Performance*. Brussels: European Commission. May.

Fangerberg, J. 2002. *Technology, Growth, and Competitiveness: Selected Essays*. Cheltenham, UK, and Northampton, MA: Edward Elgar.

Farrell, C., and M. Mandell. 1992. "Industrial Policy." *BusinessWeek*. April 4.

Featherstonhaugh, George William. 1844. *Excursion Through The Slave States, From Washington On The Potomac To The Frontier Of Mexico*. London: John Murray, Albemarle Street.

Federal Reserve of Chicago. 2007. "Can Higher Education Foster Economic Growth?—A Conference Summary. *Chicago Fed Letter*. March.

Feldman, M., and A. Link. 2001. "Innovation policy in the knowledge-based economy." In *Economics of Science, Technology and Innovation, Vol. 23*. Boston: Kluwer Academic Press.

Feldman, M., A. Link, and D. Siegel. 2002. *The Economics of Science and Technology: An Overview of Initiatives to Foster Innovation, Entrepreneurship, and Economic Growth*. Boston: Kluwer Academic Press.

Feser, E. 2005. "Industry Cluster Concepts in Innovation Policy: A Comparison of U.S. and Latin American Experience." *Interdiscliplinary Studies in Economics and Management, Vol. 4*. Vienna: Springer.

Fishback, B., C. A. Gulbranson, R. E. Litan, L. Mitchell, and M. Porzig. 2007. *Finding Business "Idols": A New Model to Accelerate Start-Ups*. Kauffman Foundation Report.

Flamm, K. 2003. "SEMATECH revisited: assessing consortium impacts on semiconductor industry R&D." In National Research Council. *Securing the Future: Regional and National Programs to Support the Semiconductor Industry*. Charles. W. Wessner, ed. Washington, DC: The National Academies Press.

Florida, R. 2002. *The Rise of the Creative Class*. New York: Basic Books.

Florida, R. 2004. "The World is Spiky." *Atlantic Monthly* October.

Fonfria, A., C. Diaz de la Guardia, and I. Alvarez. 2002. "The role of technology and competitiveness policies: a technology gap approach." *Journal of Interdisciplinary Economics* 13:223-241.

Foray, D., and P. Llerena. 1996. "Information structure and coordination in technology policy: a theoretical model and two case studies." *Journal of Evolutionary Economics* 6(2):157-173.

Fort Smith Times Record. 2010. "Mitsubishi Incentives Hit $83M." December 25.

Fox-Penner, Peter S., Marc Chupka, and Robert L. Earle. 2008. "Transforming America's Power Industry: The Investment Challenge 2010–2030." Brattle Group.

Freeman, C. 1987. *Theory of Innovation and Interactive Learning*. London: Pinter.

Friedman, T. 2005. *The World Is Flat: A Brief History of the 21st Century*. New York: W. H. Freeman.

Fry, G. R. H. 1986. "The economics of home solar water heating and the role of solar tax credits." *Land Economics* 62(2):134-144.

Fthenakis, V., J. E. Mason, and K. Zweibel. 2009. "The technical, geographical, and economic feasibility for solar energy to supply the energy needs of the US." *Energy Policy* 37(2):387-399.

Fullilove, M. T. 2005. *Root Shock: How Tearing Up City Neighborhoods Hurts America and What We Can Do About It.* New York: Ballantine Books.

Furman, J., M. Porter, and S. Stern. 2002. "The determinants of national innovative capacity." *Research Policy* 31:899-933.

Geiger, R. L., and C. M. Sá. 2009. *Tapping the Riches of Science: Universities and the Promise of Economic Growth.* Cambridge MA: Harvard University Press.

George, G., and G. Prabhu. 2003. "Developmental financial institutions as technology policy instruments: implications for innovation and entrepreneurship in emerging economies." *Research Policy* 32(1):89-108.

Grande, E. 2001. "The erosion of state capacity and European innovation policy: a comparison of German and EU information technology policies." *Research Policy* 30(6):905-921.

Grindley, P., D. Mowery, and B. Silverman. 1994. "SEMATECH and collaborative research: lessons in the design of high technology consortia." *Journal of Policy Analysis and Management* 13(4):723-758.

Grossman, G. M., and E. Helpman. 1994. "Endogenous innovation in the theory of growth." *The Journal of Economic Perspectives* 8(1):23-44.

Guidolin, M., and C. Mortarino. 2010. "Cross-country diffusion of photovoltaic systems: modelling choices and forecasts for national adoption patterns." *Technological Forecasting and Social Change* 77(2):279-296.

Gulbranson, C. A., and D. B. Audretsch. 2008. "Proof of Concept Centers: Accelerating the Commercialization of University Innovation." Ewing Marion Kauffman Foundation. January.

Hall, B. 2002. "The assessment: technology policy." *Oxford Review of Economic Policy* 18(1):1-9.

Hall, B. 2004. "University-Industry Research Partnerships in the United States." Kansai Symposium Report. February.

Hamilton, Gregory L., and Teresa A. McLendon. 2006. *Closing the Gap: An Examination and Analysis of Per Capita Personal Income in Arkansas.* August. Little Rock: University of Arkansas at Little Rock.

Harbour, K. 2011. "WV Biometrics: Fertile Ground for Innovation." Charleston, WV: West Virginia Department of Commerce.

Harris, William C. 2010. "Innovation lessons from Ireland." *Research-Technology Management* 53(1):35-39.

Hill, Edward et al. 2012. "Economic Shocks and Regional Economic Resilience." Pages 193-274 in M. Weir, N. Pindus, H. Wial and H. Wolman, eds. *Urban and Regional Policy and Its Effects, vol. 4: Building Resilient Regions.* Washington, DC: Brookings Institution Press.

Ho, Giang, and Anthony Pennington-Cross. 2005. "Fayetteville and Hot Springs Lead the Recovery in Employment." *The Regional Economist* October.

Hodges, Curt. 2011. "Beckmann Volmer Breaks Ground on Osceola Plant." *Paragould Daily Press* September 14.

Hu, Z. 2006. "IPR Policies in China: Challenges and Directions." Presentation at *Industrial Innovation in China*. Levin Institute Conference. July 24-26.

Hughes, K. 2005. *Building the Next American Century: The Past and Future of American Economic Competitiveness*. Washington, DC: Woodrow Wilson Center Press.

Hughes, K. 2005. "Facing the global competitiveness challenge." *Issues in Science and Technology* XXI(4):72-78.

Jaffe, A., J. Lerner, and S. Stern, eds. 2003. *Innovation Policy and the Economy, Vol. 3*. Cambridge, MA: MIT Press.

Janssen, M. A., R. Holahan, A. Lee, and E. Ostrom. 2010. "Lab Experiments for the Study of Social-Ecological Systems." *Science* 328(5978):613-617. April.

Jaruzelski, B., and K. Dehoff. 2008. "Beyond borders: The global innovation 1000." *Strategy and Business* 53(Winter).

Jasanoff, S., ed. 1997. *Comparative Science and Technology Policy*. Elgar Reference Collection. International Library of Comparative Public Policy, Vol. 5. Cheltenham, UK, and Lyme, NH: Edward Elgar.

Jorgenson, D., and K. Stiroh. 2002. "Raising the speed limit: economic growth in the information age." In National Research Council. 2002. *Measuring and Sustaining the New Economy*. Dale. W. Jorgenson and Charles. W. Wessner, eds. Washington, DC: The National Academies Press.

Joy, W. 2000. "Why the future does not need us." *Wired* 8(April).

Kelly, K. 1992. "Hot Spots." *BusinessWeek* October 19.

Kim, Y. 2006. "A Korean Perspective on China's Innovation System." Presentation at *Industrial Innovation in China*. Levin Institute Conference. July 24-26.

Koschatzky, K. 2003. "The regionalization of innovation policy: new options for regional change?" In G. Fuchs and P. Shapira, eds. *Rethinking Regional Innovation: Path Dependency or Regional Breakthrough?* London: Kluwer.

Krueger, A. O. "Globalization and International Locational Competition." Symposium in Honor of Herbert Giersch. Lecture delivered at the Keil Institute. May 11, 2006.

Kuhlmann, S., and J. Edler. 2003. "Scenarios of technology and innovation policies in Europe: investigating future governance—group of 3." *Technological Forecasting & Social Change* 70.

Lall, S. 2002. "Linking FDI and technology development for capacity building and strategic competitiveness." *Transnational Corporations* 11(3):39-88.

Lancaster, R. R., and M. J. Berndt. 1984. "Alternative energy development in the USA: the effectiveness of state government incentives." *Energy Policy* 12(2):170-179.

Laredo, P., and P. Mustar, eds. 2001. *Research and Innovation Policies in the New Global Economy: An International Perspective*. Cheltenham, UK: Edward Elgar.

Lee, Y. S. 2000. "The Sustainability of University-Industry Research Collaboration. *The Journal of Technology Transfer* 25(2).

Lerner, J. 1999. "Public venture capital." In National Research Council. *The Small Business Innovation Program: Challenges and Opportunities*. Charles. W. Wessner, ed. Washington, DC: National Academy Press.

Leslie, S. 1993. *The Cold War and American Science: The Military-Industrial-Academic Complex at MIT and Stanford*. New York: Columbia University Press.

Lewis, J. 2005. *Waiting for Sputnik: Basic Research and Strategic Competition*. Washington, DC: Center for Strategic and International Studies.

Lin, O. 1998. "Science and technology policy and its influence on the economic development of Taiwan." In H. S. Rowen, ed. *Behind East Asian Growth: The Political and Social Foundations of Prosperity*. London and New York: Routledge.

Link, A. N. 1995. *A Generosity of Spirit: The Early History of the Research Triangle Park*. Research Triangle Park: The Research Triangle Foundation of North Carolina.

Litan, R. E., L. Mitchell, and E. J. Reedy. 2007. "Commercializing University Innovations: Alternative Approaches." Boston: National Bureau of Economic Research. Working paper JEL No. O18, M13, 033, 034, 038.

Litan, R. E., L. Mitchell, and E. J. Reedy. 2007. "The University as Innovator: Bumps in the Road." *Issues in Science and Technology* Summer:57-66.

Lucas, R. "On the mechanics of economic development." *Journal of Military Economics* 22:38-39.

Luger, M. 2001. "Introduction: information technology and regional economic development." *Journal of Comparative Policy Analysis: Research & Practice*.

Luger, M., and H. A. Goldstein. 1991. *Technology in the Garden*. Chapel Hill: University of North Carolina Press.

Luger, M., and H. A. Goldstein. 2006. *Research Parks Redux: The Changing Landscape of the Garden*. Washington, DC: U.S. Department of Commerce.

Lundvall, B., ed. 1992. *National Innovation Systems: Towards a Theory of Innovation and Interactive Learning*. London: Pinter.

Luther, J. 2008. "Renewable Energy Development in Germany." Presentation at the NRC Christine Mirzayan Fellows Seminar. March 5, 2008. Washington, DC.

Maddison, A., and D. Johnston. 2001. *The World Economy: A Millennial Perspective*. Paris: Organization for Economic Co-operation and Development.

Mani, S. 2004. "Government, innovation and technology policy: an international comparative analysis." *International Journal of Technology and Globalization* 1(1).

Manufacturers' News. 2011. "Industrial Jobs in Arkansas Declined 1.5% Over Last Year." October 31.

Marshall, A. 1890. *Principles of Economics*. London: MacMillan & Company.

McKibben, W. 2003. *Enough: Staying Human in an Engineered Age*. New York: Henry Holt & Co.

McKinsey and Company. 2010 "Energy Efficiency, A Compelling Global Resource." McKinsey and Company.

Melissaratos, A., and N. J. Slabbert. 2009. *Innovation: The Key to Prosperity—Technology and America's Role in the 21st Century Global Economy.* Washington, DC: Montagu House.

Mendonca, M. 2007. *Feed-in Tariffs: Accelerating the Development of Renewable Energy*. London: Earthscan.

Meyer-Krahmer, F. 2001. "Industrial innovation and sustainability—conflicts and coherence." In D. Archibugi and B. Lundvall, eds. *The Globalizing Learning Economy*. New York: Oxford University Press.

Meyer-Krahmer, F. 2001. "The German innovation system." In P. Larédo and P. Mustar, eds. *Research and Innovation Policies in the New Global Economy: An International Comparative Analysis*. Cheltenham, UK: Edward Elgar.

Mills, K. G., E. B. Reynolds, and A. Reamer. 2008. *Clusters and Competitiveness: A New Federal Role for Stimulating Regional Economies.* Washington, DC: Brookings.

Moore, G. 2003. "The SEMATECH contribution." In National Research Council. *Securing the Future: Regional and National Programs to Support the Semiconductor Industry*. Charles. W. Wessner, ed. Washington, DC: The National Academies Press.

Moselle, B., J. Padilla, and R. Schmalensee. 2010. *Harnessing Renewable Energy in Electric Power Systems: Theory, Practice Policy*. Washington, DC: RFF Press.

Mufson, S. 2009. "Asian nations could outpace U.S. in developing clean energy." *Washington Post* July 16.

Murphy, L. M., and P. L. Edwards. 2003. *Bridging the Valley of Death: Transitioning from Public to Private Sector Financing*. Golden, CO: National Renewable Energy Laboratory. May.

Mustar, P., and P. Laredo. 2002. "Innovation and research policy in France (1980-2000) or the disappearance of the Colbertist state." *Research Policy* 31:55-72.

National Academy of Engineering. 2004. *The Engineer of 2020: Visions of Engineering in the New Century*. Washington, DC: The National Academies Press.

National Academy of Engineering. 2008. *Grand Challenges for Engineering.* Washington, DC: The National Academies Press.

National Academy of Sciences. 2010. *Electricity from Renewable Sources: Status, Prospects, and Impediments*. Washington, DC: The National Academies Press.

National Academy of Sciences, National Academy of Engineering, and Institute of Medicine. 2007. *Rising Above the Gathering Storm: Energizing and Employing America for a Brighter Economic Future*. Washington, DC: The National Academies Press.

National Academy of Sciences, National Academy of Engineering, and National Research Council. 2009. *America's Energy Future: Technology and Transformation*. Washington, DC: The National Academies Press.

National Academy of Sciences, National Academy of Engineering, and National Research Council. 2009. *Real Prospects for Energy Efficiency in the United States*. Washington, DC: The National Academies Press.

National Economic Council and Office of Science and Technology Policy. 2009. "A Strategy for American Innovation: Driving Towards Sustainable Growth and Quality Jobs." Washington, DC: Executive Office of the President. September.

National Governors' Association. 2007. *Innovation America.* Washington, DC: National Governors' Association.

National Institute of Standards and Technology. 2010. "NIST Manufacturing Extension Partnership Awards $9.1 Million for 22 Projects to Enhance U.S. Manufacturers' Global Competitiveness." Gaithersburg, MD: National Institute of Standards and Technology. October 5.

National Research Council. 1996. *Conflict and Cooperation in National Competition for High-Technology Industry*. Washington, DC: National Academy Press.

National Research Council. 1999. *The Advanced Technology Program: Challenges and Opportunities.* Charles. W. Wessner, ed. Washington, DC: National Academy Press.

National Research Council. 1999. *Funding a Revolution: Government Support for Computing Research*. Washington, DC: National Academy Press.

National Research Council. 1999. *Industry-Laboratory Partnerships: A Review of the Sandia Science and Technology Park Initiative.* Charles. W. Wessner, ed. Washington, DC: National Academy Press.

National Research Council. 1999. *New Vistas in Transatlantic Science and Technology Cooperation.* Charles. W. Wessner, ed. Washington, DC: National Academy Press.

National Research Council. 1999. *The Small Business Innovation Research Program: Challenges and Opportunities.* Charles. W. Wessner, ed. Washington, DC: National Academy Press.

National Research Council. 1999. *U.S. Industry in 2000: Studies in Competitive Performance*. D. C. Mowery, ed. Washington, DC: National Academy Press.

National Research Council. 2000. *The Small Business Innovation Research Program: A Review of the Department of Defense Fast Track Initiative.* Charles. W. Wessner, ed. Washington, DC: National Academy Press.

National Research Council. 2001. *A Review of the New Initiatives at the NASA Ames Research Center*. Charles. W. Wessner, ed. Washington, DC: National Academy Press.

National Research Council. 2001. *Building a Workforce for the Information Economy.* Washington, DC: National Academy Press.

National Research Council. 2001. *Capitalizing on New Needs and New Opportunities: Government-Industry Partnerships in Biotechnology and Information Technologies*. Charles. W. Wessner, ed. Washington, DC: National Academy Press.

National Research Council. 2001. *The Advanced Technology Program: Assessing Outcomes*. Charles. W. Wessner, ed. Washington, DC: National Academy Press.

National Research Council. 2001. *Trends in Federal Support of Research and Graduate Education.* S. A. Merrill, ed. Washington, DC: National Academy Press.

National Research Council. 2002. *Partnerships for Solid-State Lighting.* Charles. W. Wessner, ed. Washington, DC: National Academy Press.

National Research Council. 2003. *Government-Industry Partnerships for the Development of New Technologies: Summary Report.* Charles. W. Wessner, ed. Washington, DC: The National Academies Press.

National Research Council. 2003. *Securing the Future: Regional and National Programs to Support the Semiconductor Industry*. Charles. W. Wessner, ed. Washington, DC: The National Academies Press.

National Research Council. 2003. *Understanding Climate Change Feedbacks*. Washington, DC: The National Academies Press.

National Research Council. 2004. *Productivity and Cyclicality in Semiconductors: Trends, Implications, and Questions*. Dale W. Jorgenson and Charles. W. Wessner, eds. Washington, DC: The National Academies Press.

National Research Council. 2004. *The Small Business Innovation Research Program: Program Diversity and Assessment Challenges.* Charles. W. Wessner, ed. Washington, DC: The National Academies Press.

National Research Council. 2005. *Deconstructing the Computer*. Dale W. Jorgenson and Charles. W. Wessner, eds. Washington, DC: The National Academies Press.

National Research Council. 2005. *Getting Up to Speed: The Future of Superconducting*. S. L. Graham, M. Snir, and C. A. Patterson, eds. Washington, DC: The National Academies Press.

National Research Council. 2005. *Policy Implications of International Graduate Students and Postdoctoral Scholars in the United States*. Washington, DC: The National Academies Press.

National Research Council. 2006. *Software, Growth, and the Future of the U.S. Economy*. Dale W. Jorgenson and Charles. W. Wessner, eds. Washington, DC: The National Academies Press.

National Research Council. 2006. *The Telecommunications Challenge: Changing Technologies and Evolving Policies*. Charles. W. Wessner, ed. Washington, DC: The National Academies Press.

National Research Council. 2007. *Enhancing Productivity Growth in the Information Age: Measuring and Sustaining the New Economy*. Dale. W. Jorgenson and Charles. W. Wessner, eds. Washington, DC: The National Academies Press.

National Research Council. 2007. *Innovation Policies for the 21st Century*. Charles. W. Wessner, ed. Washington, DC: The National Academies Press.

National Research Council. 2007. *India's Changing Innovation System: Achievements, Challenges, and Opportunities for Cooperation*. Charles. W. Wessner and Sujai. J. Shivakumar, eds. Washington, DC: The National Academies Press.

National Research Council. 2007. *SBIR and the Phase III Challenge of Commercialization*. Charles. W. Wessner, ed. Washington, DC: The National Academies Press.

National Research Council. 2008. *An Assessment of the Small Business Innovation Research Program*. Charles. W. Wessner, ed. Washington, DC: The National Academies Press.

National Research Council. 2008. *An Assessment of the Small Business Innovation Research Program at the Department of Energy*. Charles. W. Wessner, ed. Washington, DC: The National Academies Press.

National Research Council. 2008. *An Assessment of the Small Business Innovation Research Program at the National Science Foundation*. Charles. W. Wessner, ed. Washington, DC: The National Academies Press.

National Research Council. 2008. *Innovative Flanders: Innovation Policies for the 21st Century*. Charles. W. Wessner, ed. Washington, DC: The National Academies Press.

National Research Council. 2008. *Innovation in Global Industries: U.S. Firms Competing in a New World*. J. Macher and D. Mowery, eds. Washington, DC: The National Academies Press.

National Research Council. 2008. *The National Academies Summit on America's Energy Future: Summary of a Meeting*. Washington, DC: The National Academies Press.

National Research Council. 2009. *21st Century Innovation Systems for Japan and the United States: Lessons from a Decade of Change*. S. Nagaoka, M. Kondo, K. Flamm, and C. Wessner, eds. Washington, DC: The National Academies Press.

National Research Council. 2009. *An Assessment of the Small Business Innovation Research Program at the Department of Defense*. Charles. W. Wessner, ed. Washington, DC: The National Academies Press.

National Research Council. 2009. *An Assessment of the Small Business Innovation Research Program at the National Aeronautics and Space Administration*. Charles. W. Wessner, ed. Washington, DC: The National Academies Press.

National Research Council. 2009. *An Assessment of the Small Business Innovation Research Program at the National Institutes of Health*. Charles. W. Wessner, ed. Washington, DC: The National Academies Press.

National Research Council. 2009. *Hidden Costs of Energy: Unpriced Consequences of Energy Production and Use*. Washington, DC: The National Academies Press.

National Research Council. 2009. *Revisiting the Department of Defense SBIR Fast Track Initiative*. Charles. W. Wessner, ed. Washington, DC: The National Academies Press.

National Research Council. 2009. *Understanding Research, Science and Technology Parks: Global Best Practices*. Charles. W. Wessner, ed. Washington, DC: The National Academies Press.

National Research Council. 2009. *Venture Funding and the NIH SBIR Program*. Charles. W. Wessner, ed. Washington, DC: The National Academies Press.

National Research Council. 2010. *Managing University Intellectual Property in the Public Interest*. Stephen Merrill and Anne-Marie Mazza, eds., Washington, DC: The National Academies Press.

National Research Council. 2011. *Building the 21st Century: U.S.-China Cooperation on Science, Technology, and Innovation*. Charles. W. Wessner, rapporteur. Washington, DC: The National Academies Press.

National Research Council. 2011. *Growing Innovation Clusters for American Prosperity*, Charles W. Wessner, rapporteur, Washington, DC: The National Academies Press.

National Research Council. 2011. *The Future of Photovoltaics Manufacturing in the United States.* Charles. W. Wessner, rapporteur. Washington, DC: The National Academies Press.

National Research Council. 2012. *Building Hawaii's Innovation Economy*. Charles. W. Wessner, rapporteur. Washington, DC: The National Academies Press.

National Research Council. 2012. *Clustering for 21st Century Prosperity*. Charles. W. Wessner, rapporteur. Washington, DC: The National Academies Press.

National Research Council. 2012. *Meeting Global Challenges: German-U.S. Innovation Policy*. Charles. W. Wessner, rapporteur. Washington, DC: The National Academies Press.

National Research Council. 2012. *Rising to the Challenge: U.S. Innovation Policy for the Global Economy*. Charles. W. Wessner and Alan Wm. Wolff, editors. Washington, DC: The National Academies Press.

National Research Council of Canada. 2008. *State of the Nation 2008: Canada's Science, Technology, and Innovation System*. Ottawa: Government of Canada.

National Science Board. 2010. *Science and Engineering Indicators 2010*. Arlington, VA: National Science Foundation.

Needham, J. 1954-1986. *Science and Civilization in China* (five volumes). Cambridge: Cambridge University Press.

Nelson, R., and K. Nelson. 2002. "Technology, institutions, and innovation systems." *Research Policy* 31:265-272.

Nelson, R., and N. Rosenberg. 1993. "Technical innovation and national systems." In R. R. Nelson, ed. *National Innovation Systems: A Comparative Analysis*. Oxford: Oxford University Press.

North American Windpower. 2010. "Mitsubishi Breaks Ground on Nacelle Facility in Arkansas." October 8.

NWA Online. 2010. "Firm Building Jonesboro Plant to Get $22 Million Stimulus." January 11.

O'Hara, M. P. 2005. *Cities of Knowledge: Cold War Science and the Search for the Next Silicon Valley.* Princeton: Princeton University Press.

O'Reilly, Joseph. 2009. "Arkansas: A Natural Wonder." *Inbound Logistics* May.

Organisation for Economic Co-operation and Development. 1997. "National Innovation Systems." Paris: Organisation for Economic Co-operation and Development.

Organisation for Economic Co-operation and Development. 2009. *Main Science and Technology Indicators*. Paris: Organisation for Economic Co-operation and Development.

Orszag, P., and T. Kane. 2003. "Funding Restrictions at Public Universities: Effects and Policy Implications." *Brookings Institution Working Paper*. September.

Oughton, C., M. Landabaso, and K. Morgan. 2002. "The regional innovation paradox: innovation policy and industrial policy." *Journal of Technology Transfer* 27(1).

Palmintera, D. 2005. *Accelerating Economic Development through University Technology Transfer*. Reston, VA: Innovation Associates.

Pavitt, K. 1998. "The Social Shaping of the National Science Base." *Research Policy* 27:793-805.

Pezzini, M. 2003. "Cultivating Regional Development: Main Trends and Policy Challenges in OECD Regions." Paris: Organisation for Economic Co-operation and Development.

Plastics News. 2011. "River Bend Gets Kosmo Work." November 7.

Porter, M. E. 1990. *The Competitive Advantage of Nations.* New York: The Free Press.

Porter, M. E., ed. 1993. *Choosing to Compete: A Statewide Strategy for Job Creation and Economic Growth*. Boston: The Commonwealth of Massachusetts.

Porter, M. E. 1998. "Clusters and the new economics of competition" *Harvard Business Review* 76(6):77-90.

Porter, M. E. 2005. *Clusters of Innovation Initiative: Regional Foundations of U.S. Competitiveness*. Washington DC: Council on Competitiveness.

Posen, A. 2001. "Japan." In B. Steil, D. G. Victor, and R. R. Nelson, eds. *Technological Innovation and Economic Performance*. Princeton: Princeton University Press.

President's Council of Advisors on Science and Technology. 2004. *Sustaining the Nation's Innovation System: Report on Information Technology Manufacturing and Competitiveness*. Washington, DC: Executive Office of the President. January.

PricewaterhouseCoopers. 2006. "China's Impact on the Semiconductor Industry: 2005 Update." PricewaterhouseCoopers.

PricewaterhouseCoopers and National Venture Capital Association. 2010. "MoneyTree Report." PricewaterhouseCoopers.

Pulaski County Chancery Court. 2001. *Lake View School District No. 25 v. Huckabee*. No. 1992-5318. May 25.

Purdue University. 2009. *Crossing the Next Regional Frontier: Information and Analytics Linking Regional Competitiveness to Investment in a Knowledge-Based Economy*. West Lafayette, IN: Purdue University. October.

Raduchel, W. 2006. "The end of stovepiping." In National Research Council. *The Telecommunications Challenge: Changing Technologies and Evolving Policies*, Charles. W. Wessner, ed., Washington, DC: The National Academies Press.

Ragwitz, M., and C. Huber. 2005. "Feed-in systems in Germany and Spain: a comparison." Fraunhofer Institut für Systemtechnik und Innovationsforschung.

Reid, T. R. 2004. *The United States of Europe: The New Superpower and the End of American Supremacy*. New York: Penguin Press.

Renewable Energy Policy Network for the 21st Century. 2009. *Renewables Global Status Report 2009*. Paris: REN21.

Rickerson, W., and R. Grace. 2007. "The Debate Over Fixed Price Incentives for Renewable Electricity in Europe and the United States: Fallout and Future Directions." White Paper prepared for the Heinrich Böll Foundation. Washington, DC.

Rolnick, Arthur J. 2007. "Congress Should End the Economic War Among the States." Testimony before the House Domestic Policy Subcommittee. October 10.

Romer, P. M. 1990. "Endogenous technological change." *Journal of Political Economy* October.

Rosenberg, N., and R. R. Nelson. 1994. "American universities and technical advance in industry." *Research Policy* 23:323-248

Ruttan, V. 2002. *Technology, Growth and Development: An Induced Innovation Perspective*. Oxford: Oxford University Press.

Rutten, R., and F. Boekema. 2005. "Innovation, policy and economic growth: theory and cases." *European Planning Studies* 13(8).

Sallet, J., E. Paisley, and J. R. Masterman. 2009. "The Geography of Innovation: the Federal Government and the Growth of Regional Innovation clusters." *Science Progress.* September.

Sarzynski, A. 2010. "The Impact of Solar Incentive Programs in Ten States." George Washington Institute of Public Policy Technical Report. Revised March 2010.

Saxenian, A. 1994. *Regional Advantage: Culture and Competition in Silicon Valley and Route 128.* Cambridge, MA: Harvard University Press.

Schumpeter, Joseph A. 1975. *Capitalism, Socialism, and Democracy.* New York: Harper.

Scott, A. J. 2004. *On Hollywood: The Place, the Industry.* Princeton NJ: Princeton University Press.

SERI (Solar Energy Industries Association). 2009. *U.S. Solar in Review 2008.* Washington, DC: Solar Energy Industries Association.

Shang, Y. 2006. "Innovation: New National Strategy of China." Presentation at Industrial Innovation in China. Levin Institute Conference. July 24-26.

Sheehan, J., and A. Wyckoff. 2003. "Targeting R&D: Economic and Policy Implications of Increasing R&D Spending." DSTI/DOC(2003)8. Paris: Organisation for Economic Co-operation and Development.

Sherwood, L. 2008. *U.S. Solar Market Trends 2007.* Latham, NY: Interstate Revewable Energy Council.

Small Business Administration. 2010. "SBA Announces Support for 10 Regional 'Innovative Economies' Clusters, Local Job Creation." SBA News Release 10-50. October 20.

Smits, R., and S. Kuhlmann. 2004. "The rise of systemic instruments in innovation policy." *International Journal of Foresight and Innovation Policy* 1(1/2).

Sparks, Glen R. 2007. "Community Profile: Conway, Ark. Makes Play for Economic Boom." *The Regional Economist* July.

Speck, S. 2008. "The design of carbon and broad-based energy taxes in European countries." *Vermont Journal of Environmental Law* 10.

Spencer, W., and T. E. Seidel. 2004. "International technology roadmaps: The U.S. semiconductor experience." In National Research Council. *Productivity and Cyclicality in Semiconductors: Trends, Implications, and Questions.* D. W. Jorgenson and Charles W. Wessner, eds. Washington, DC: The National Academies Press.

Stanford University. 1999. *Inventions, Patents and Licensing: Research Policy Handbook.* Document 5.1. July 15.

Stokes, D. E. 1997. *Pasteur's Quadrant: Basic Science and Technological Innovation.* Washington, DC: Brookings Institution.

Sturgeon, T. J. 2000. "How Silicon Valley Came to Be." In M. Kenney (ed.), *Understanding Silicon Valley: The Anatomy of an Entrepreneurial Region* (pp. 15-47). Stanford, CA: Stanford University Press.

Swamidass, P. M., and V. Vulasa. 2009. "Why university inventions rarely produce income? Bottlenecks in university technology transfer." *The Journal of Technology Transfer* 34(4).

Taleb, N. N. 2007. *The Black Swan: The Impact of the Highly Improbable.* New York: Random House.

Tan, J. 2006. "Growth of industry clusters and innovation: lessons from Beijing Zhongguancun Science Park." *Journal of Business Venturing* 21(6):827-850. November.

Task Force for the Creation of Knowledge-Based Jobs. 2002. *Report of the Task Force for the Creation of Knowledge-Based Jobs.* Accessed at *<http://www.asta.arkansas.gov/resources/Documents/Knowledge-Based%20Jobs%20Report.pdf>*. September.

Tassey, G. 2004. "Policy issues for R&D investment in a knowledge-based economy." *Journal of Technology Transfer* 29:153-185.

Taylor, M. 2008. "Beyond technology-push and demand-pull: lessons from California's solar policy." *Energy Economics* 30(6):2829-2854.

Teubal, M. 2002. "What is the systems perspective to innovation and technology policy and how can we apply it to developing and newly industrialized economies?" *Journal of Evolutionary Economics* 12(1-2).

Thomas, Kenneth P. 2011. *Investment Incentives and the Global Competition for Capital.* London and Basingstoke: Palgrave MacMillan.

Thompson, Susan C. 2010. "Factory Closing Shocks Community into Opening Wallets for Economic Development." *The Regional Economist.* October.

Tödtling, F., and M. Trippl. 2005 "One size fits all? Towards a differentiated regional innovation policy approach." *Research Policy* 34.

Tol, R. S. J. 2008. "The social cost of carbon: trends, outliers, and catastrophes." *Economics—the Open-Access, Open-Assessment E-Journal* 2(25):1-24.

Tzang, C. 2010. "Managing innovation for economic development in greater China: The origins of Hsinchu and Zhongguancun." *Technology in Society* 32(2):110-121. May.

University of Arkansas College of Engineering. 2008. "University of Arkansas Installing Supercomputer, 'Star of Arkansas', to be State's Fastest." Press Release. Fayetteville, AR: University of Arkansas at Fayetteville.

U.S. Department of Energy. 2006. Press Release. "Department Requests $4.1 Billion Investment as Part of the American Competitiveness Initiative: Funding to Support Basic Scientific Research." February 2.

U.S. General Accounting Office. 2002. *Export Controls: Rapid Advances in China's Semiconductor Industry Underscore need for Fundamental U.S. Policy Review*. GAO-020620. Washington, DC: U.S. General Accounting Office. April.

Van Looy, B., K. Debackere, and T. Magerman. 2005. *Assessing Academic Patent Activity: The Case of Flanders.* Leuven: SOOS.

Van Looy, B., M. Ranga, J. Callaert, K. Debackere, and E. Zimmermann. 2004. "Combining Entrepreneurial and Scientific Performance in Academia: Towards a Compounded and Reciprocal Matthew-effect?" *Research Policy* 33(3):425-441.

Veugelers, R., J. Larosse, M. Cincera, D. Carchon, and R. Kalenga-Mpala. 2004. "R&D activities of the business sector in Flanders: results of the R&D surveys in the context of the 3% target." Brussels: IWT-Studies.

Wang, C. 2005. "IPR sails against current stream." *Caijing* October 17.

Wang, Q. 2010. "Effective policies for renewable energy—the example of China's wind power—lessons for China's photovoltaic power." *Renewable and Sustainable Energy Reviews* 14(2):702-712.

Wessner, C. W. 2005. "Entrepreneurship and the innovation ecosystem." In D. B. Audretsch, H. Grimm, and C. W. Wessner, eds. *Local Heroes in the Global Village: Globalization and the New Entrepreneurship Policies*. New York: Springer.

Wessner, C. W. 2005. *Partnering Against Terrorism*. Washington, DC: The National Academies Press.

Wind Power News. 2011. "A Wind Study the Size of Arkansas." April 1.

Wind Systems. 2011. Interview with Joe Brenner, Vice President of Nordex USA. January.

Wiser, R., G. Barbose, C. Peterman, and N. Darghouth. 2009. *Tracking the Sun II: The Installed Cost of Photovoltaics in the U.S. from 1998-2008*. Berkeley, CA: Lawrence Berkeley National Laboratory.

Witt, C. E., R. L. Mitchell, and G. D. Mooney. 1993. "Overview of the Photovoltaic Manufacturing Technology (PVMaT) Project." Paper presented at the 1993 National Health Transfer Conference. August 8-11. Atlanta, Georgia.

Xi, Lu Michael B. McElroy, and Juha Kiviluoma. 2009. "Global potential for wind-generated electricity." *Proceedings of the National Academy of Sciences of the United States of America* 106(27):10933-10938.

Yu, J., and R. Jackson. 2011. "Regional Innovation Clusters: A Critical Review." *Growth and Change* 42(2).

Zeigler, N. 1997. *Governing Ideas: Strategies for Innovation in France and Germany*. Ithaca, NY, and London: Cornell University Press.

Zweibel, K. 2010. "Should solar photovoltaics be deployed sooner because of long operating life at low, predictable cost?" *Energy Policy* 38(11):7519-7530.